Computer Simulation of Aerial Target Radar Scattering, Recognition, Detection, and Tracking

For a complete listing of the *Artech House Radar Library*,
turn to the back of this book.

Computer Simulation of Aerial Target Radar Scattering, Recognition, Detection, and Tracking

Yakov D. Shirman
Editor

Artech House
Boston • London
www.artechhouse.com

Library of Congress Cataloging-in-Publication Data

A catalog record for this book is available from the Library of Congress.

British Library Cataloguing in Publication Data

A catalog record for this book is available from the British Library.

Cover design by Igor Valdman

International Standard Book Number: 1-58053-172-5

10 9 8 7 6 5 4 3 2 1

Contents

Preface

Scattering of radar targets has become one of the most important parts of modern radar system analysis [1–3]. Computer simulation of the radar targets' scattering is of great importance in the initial research and development (R&D) steps of recognition, detection, and tracking, and in modern radar education [4–7]. Since aerial target orientation can be estimated only approximately, the statistics of scattered signals are more important than their exact values. This allows the wide use of scattering theory approximations.

Radar recognition development requires expensive experiments. The task of simulation in recognition is to replace such experiments in initial R&D steps. As an experiment, the simulation [4–7] permits the choice of various recognition performances:

- Alphabets of target classes or types recognized;
- Recognition features (signatures) and their combinations;
- Illuminating signals and decision rules;
- Radar Subsystems Tolerances.

Simulation programs created for recognition can be used easily for detection and tracking. As for detection, the critical discussion [8–10] of classical Swerling backscattering statistics has shown an intention to look for new models [10, p. 718]. As for tracking, essential factors of secondary modulation and other nonstationary scattering factors [1, 11] also lead to the use of new models.

The peculiarity of simulation consists of taking into account:

- The variety of radar targets, their orientations, positions, and features of their rotating parts;
- The electrodynamics of backscattering for every target in any orientation;
- The variety of illuminating signals, their space-time-polarization transform in the scattering and posterior processing;
- The statistics of target motion in real atmosphere.

Chapters 1 through 6 of the book, which describe monostatic radar and "nonstealth" targets, consider:

- The foundations of scattering simulation on centimeter and decimeter waves given in the simplest components approximation (Chapter 1);
- Review and simulation of recognition features (signatures) for wideband and narrowband illumination of targets (Chapters 2 and 3);
- Review and simulation of recognition algorithms' operation (Chapter 4);
- Peculiarity of backscattering simulation and recognition for low-altitude targets (Chapter 5);
- Review and simulation of signals' detection and operation of simplest algorithms of target tracking (Chapter 6).

Chapters 1 through 4 and Chapter 6 are supplemented by a CD-ROM program disc and manual for practical simulation, which are now both available as a possible instrument for R&D.

Chapter 7 attempts to expand the simulation possibilities for the case of targets with reduced cross section and covered with radar-absorbing materials and for the case of bistatic radar [2, 12, 13]. The augmented physical optics approximation, mostly without simplest components introduction, is used here.

This book is the first variant of the topic to discuss all-round computer simulation of real targets' secondary radiation, serving to solve recognition problems primarily. Joint consideration of various recognition algorithms and their operation, which is absent in the most recent technical literature, may also be of interest. We expect that in the future the backscattering simulation programs will be systematically improved on the basis of new experimental and physical simulation data. Our current simulation programs are a definite step in this direction.

The text of the book appears as the outcome of the authors' group work: S.A. Gorshkov contributed to Chapters 1 through 5; S. P. Leshchenko contributed to Chapters 1 through 4 and Chapter 6; V. M. Orlenko contributed to all chapters and helped edit the book; S. Yu. Sedyshev contributed to Chapter 5; O. I. Sukharevsky contributed to Chapter 7; Ya. D. Shirman contributed to all chapters and edited the book.

Great support and help were rendered to the authors by the Radar Series Editor, Professor David K. Barton, who attentively read the entire text, gave valuable advice to the authors, and corrected their nonnative English. We are very grateful to him.

References

[1] Barton, D. K., *Modern Radar System Analysis*, Norwood, MA: Artech House, 1988.

[2] Knott, E. F., J. F. Shaefer, and M. T. Tuley, *Radar Cross Section*, Second Edition, Norwood, MA: Artech House, 1993.

[3] Rihaczek, A. W., and S. I. Hershkowitz, *Radar Resolution and Complex-Image Analysis*, Norwood, MA: Artech House, 1996.

[4] Shirman, Y. D., et al., "Aerial Target Backscattering Simulation and Study of Radar Recognition, Detection and Tracking," *IEEE Int. Radar-2000*, Washington, DC, May 2000, pp. 521–526.

[5] Shirman, Y. D., et al., "Study of Aerial Target Radar Recognition by Method of Backscattering Computer Simulation," *Proc. Antenna Applications Symp.*, September 1999, Allerton Park Monticello, Illinois, pp. 431–447.

[6] Shirman, Y. D., et al., "Methods of Radar Recognition and Their Simulation," *Zarubeghnaya Radioelectronika-Uspehi Sovremennoi Radioelectroniki*, November 1996, Moscow, pp. 3–63; and *Collection of Papers*, Issue 3, 2000, Moscow: Radiotechnika Publishing House, pp. 5–64 (in Russian).

[7] Gorshkov, S. A, "Experimental and Computational Methods of Secondary Radiation Performance Evaluation." In *Handbook: Electronic Systems: Construction Foundations and Theory*, pp. 163–179, Y. D. Shirman (ed.), Moscow: Makvis Publishing House, 1998 (in Russian).

[8] Swerling, P., "Radar Probability of Detection for Some Additional Fluctuating Target Cases," *IEEE Trans.*, AES-33, No. 2, Part 2, April 1997, pp. 698–709.

[9] Xu, X., and P. Huang, "A New RCS Statistical Model of Radar Targets," *IEEE Trans.*, AES-33, No. 2, Part 2, April 1997, pp. 710–714.

[10] Johnston, S. L., "Target Model Pitfalls (Illness, Diagnosis, and Prescription)," *IEEE Trans.*, AES-33, No. 2, Part 2, April 1997, pp. 715–720.

[11]   Ostrovityanov, R. V., and F. A. Basalov, *Statistical Theory of Extended Radar Targets*, Moscow: Soviet Radio Publishing House, 1982; Norwood, MA: Artech House, 1985.

[12]   Ufimtsev, P. Y., "Comments on Diffraction Principles and Limitations of RCS Techniques," *Proc. IEEE* 84, April 1996.

[13]   Sukharevsky, O. I., et al., "Calculation of Electromagnetic Wave Scattering on Perfectly Conducting Object Partly Coated by Radar Absorbing Material with the Use of Triangulation Cubature Formula," *Radiophyzika and Radioastronomiya*, Vol. 5, No. 1, 2000 (in Russian).

1

Foundations of Scattering Simulation on Centimeter and Decimeter Waves

In the beginning of this chapter we consider initial information about target scattering (Section 1.1) and compare the analog and digital computer methods of scattering simulation (Sections 1.2 and 1.3). For simulation in centimeter and decimeter radar wavebands we select here the simplest component variant of target description and its scattering computer simulation. By introducing a set of coordinate systems, the follow-up consideration of moving targets and their elements is provided. The general equation of backscattering is given also for the far-field zone (Section 1.3). The peculiarities of backscattering simulation for deterministic and random target motion are considered in Section 1.4. Simulation peculiarities of backscattering from the targets' rotating parts are considered in Section 1.5. Simulated radar quality (performance) indices are discussed in Section 1.6.

Comparison of the simulation results with experimental ones is an important but complicated task. Such a comparison will be carried out mostly in Chapters 2 through 6 in connection with the peculiarities of information being received, but we will note such a comparison in Chapter 1 also.

1.1 Target Scattering

We consider in this section the scattering phenomenon and its main radar characteristics. In connection with the spreading of extended broadband signals, the doppler effect is considered not only as a change in the signal's carrier frequency but as its whole time-frequency scale transform (doppler transform).

1.1.1 Scattering Phenomenon and Its Main Radar Characteristics

The scattering phenomenon arises when arbitrary waves illuminate an obstacle. Any heterogeneity of electric and magnetic parameters of a propagation medium serves as an obstacle for radio waves. The incident wave excites oscillations in the obstacle that are the origins of secondary radiation (scattering) in various directions. Especially important for the widely used monostatic radar is the scattering in the direction opposite to the direction of the incident wave propagation (the so-called backscattering). The character of such scattering (backscattering) depends on the target's material, size, configuration, wavelength, modulation law, polarization, and the specifics of target trajectory and motion of its internal elements.

As usual, we'll consider only linear scattering. The most important radar characteristics are the *radar cross-section* (RCS or σ_{tg}) and the *polarization scattering matrix*.

RCS is the most important characteristic of a target independent of the distance R from radar to target. Let us begin its consideration with the case where the resolution cell embraces the whole target.

The IEEE Dictionary defines the RCS formally as a measure of reflective strength defined as 4π times the ratio of power density Π'_{rec} per unit solid angle (in watts per steradian) scattered in the direction of the receiver to the power density Π_{tg} per unit area (in watts per square meter) in a plane wave incident on the scatterer (target) from the direction of the transmitter:

$$\sigma_{tg} = \lim_{R\to\infty} 4\pi \frac{\Pi'_{rec}}{\Pi_{tg}} = \lim_{R\to\infty} 4\pi R^2 \frac{\Pi_{rec}}{\Pi_{tg}} = \lim_{R\to\infty} 4\pi R^2 \frac{|E_{rec}|^2}{|E_{tg}|^2} \quad (1.1)$$

Here, $\Pi_{rec} = \Pi'_{rec}/R^2$ is the power density per unit area (in watts per square meter) at the receiver, E_{rec} is the electric field magnitude at the receiver, and E_{tg} is the electric field magnitude incident on the target.

The formal definition (1.1) can be explained by replacing the distant real target with the equivalent isotropic scatterer without losses so that:

1. It produces in the direction of the radar receiver antenna the same power density Π_{rec} (in watts per square meter) as the real target does;

2. It intercepts, as it is supposed to, a power $\sigma_{tg}\Pi_{tg}$ from the power flux near the target with the density Π_{tg} (in watts per square meter).

In the assumed condition of isotropic scattering, the scattered power is distributed uniformly through the surface area $4\pi R^2$ of a sphere centered on the target. In the absence of power losses one has the equations

$$\sigma_{tg}\Pi_{tg} = 4\pi R^2 \Pi_{rec}, \qquad \sigma_{tg} = 4\pi R^2 \frac{\Pi_{rec}}{\Pi_{tg}}$$

equivalent to (1.1).

If the polarization characteristics of illuminating wave k (linear, circular, elliptic) and that of receiver antenna l differ from one another, the value of $\sigma_{tg} = \sigma_{k,l}$ depends on these polarization characteristics k and l.

For the bistatic radar the value of $\sigma_{k,l}$ depends on direction angle vectors of incident $\boldsymbol{\theta}_1 = \|\beta_1 \quad \epsilon_1\|^T$ and of scattered to receiver $\boldsymbol{\theta}_2 = \|\beta_2 \quad \epsilon_2\|^T$ waves. The wave's power flux density is proportional to the square of its electrical field intensity $|E_l(\boldsymbol{\theta}_1, \boldsymbol{\theta}_2)|$; therefore,

$$\sigma_{k,l}(\boldsymbol{\theta}_1, \boldsymbol{\theta}_2) = 4\pi \frac{\Pi'_l(\boldsymbol{\theta}_1, \boldsymbol{\theta}_2)}{\Pi_k} = 4\pi R^2 \frac{|E_l(\boldsymbol{\theta}_1, \boldsymbol{\theta}_2)|^2}{|E_k|^2} \qquad (1.2)$$

For the monostatic radar with common transmitting and receiving antenna

$$\boldsymbol{\theta}_1 = \boldsymbol{\theta}_2 = \boldsymbol{\theta}, \qquad l = k$$

which simplifies (1.2).

For the case when the target elements are resolved, the sum of partial mean RCS will be considered as the target's mean RCS [1–6].

The polarization scattering matrix (PSM) is used in the general case of polarization transformation from an incident wave to scattered one [1–6]. The PSM has the form

$$\mathbf{A} = \left\| \sqrt{\sigma_{k,l}} \cdot e^{j\varphi_{k,l}} \right\| = \left\| \begin{matrix} \sqrt{\sigma_{11}} \cdot e^{j\varphi_{11}} & \sqrt{\sigma_{12}} \cdot e^{j\varphi_{12}} \\ \sqrt{\sigma_{21}} \cdot e^{j\varphi_{21}} & \sqrt{\sigma_{22}} \cdot e^{j\varphi_{22}} \end{matrix} \right\| \qquad (1.3)$$

where for monostatic radar $\sigma_{21} = \sigma_{12}$ and $\varphi_{21} = -\varphi_{12}$. To obtain the PSM it is necessary to introduce the *polarization basis* consisting of two polarized waves of orthogonal polarization, elliptical in general. Typical bases are those of horizontal and vertical linear polarizations, of two circular polarizations with opposite rotating directions, and the target's own polarization basis, which will be considered below. Values of $\sigma_{k,l}$ are determined by (1.2). Values of indices k, l = 1 correspond to the first type of wave polarization, and k, l = 2 correspond to the second one. Values of $\varphi_{k,l}$ characterize phase shifts due to wave propagation on the radar-target-radar trace.

The PSM in general is a nondiagonal matrix with five independent parameters σ_{11}, σ_{22}, $\sigma_{12} = \sigma_{21}$, φ_{22}, $-\varphi_{11}$, $\varphi_{12} - \varphi_{11} = -(\varphi_{21} - \varphi_{11})$ for monostatic radar. Parameters of the nondiagonal elements σ_{12}, σ_{21}, $\varphi_{12} - \varphi_{11}$, $\varphi_{21} - \varphi_{11}$ characterize the polarization transformation in the scattering process. If the PSM is diagonal and $\sigma_{11} = \sigma_{22}$, then polarization transformation is absent. It corresponds, for example, to reflection from a conducting sphere. Knowing the PSM, one can express the electrical field intensity vector near the receiver $\mathbf{E}_{rec} = \mathbf{AE}_{tg}/\sqrt{4\pi R^2}$ through that near the target $\mathbf{E}_{tg}$.

Normalized Antennas Polarization Vector.   In an antenna's transmitting mode such a vector

$$\mathbf{p}^0 = \left\| \cos\gamma \quad e^{j\delta} \cdot \sin\gamma \right\|^T \tag{1.4}$$

determines the vector of electrical field components of regularly polarized transmitted wave

$$\mathbf{E}(t) = \mathrm{Re}[\dot{E} \cdot \mathbf{p}^0 e^{j2\pi f_0 t}]$$

$$= \left\| E_1 \cos(2\pi f_0 t - \psi_1) \quad E_2 \cos(2\pi f_0 t - \psi_2) \right\|^T$$

Vector $\mathbf{E}$ is supposed to have components along some mutually orthogonal unit vectors $\mathbf{l}^0$, $\mathbf{m}^0$, each of them being orthogonal to the propagation unit vector of the incident wave. In the presented equations the values of E, $\cos\gamma$, $\sin\gamma$, and δ are determined as

$$E = |\dot{E}| \cdot e^{-j\psi_1}, \quad |\dot{E}| = \sqrt{E_1^2 + E_2^2}, \quad \cos\gamma = E_1/|E|,$$

$$\sin\gamma = E_2/|E|, \quad \delta = \psi_1 - \psi_2$$

The condition of an antenna's polarization vector normalization $(\mathbf{p}^0)^{*T} \cdot \mathbf{p}^0 = 1$ is fulfilled. Elliptical polarization degenerates into the linear one for $\delta = 0$ and into the circular one for $\delta = \pm\pi/2$, $\gamma = \pi/4$. Due to the antenna reciprocity principle, the vector $\mathbf{p}^0$, (1.4), can be used for a receiving antenna also. Introduction of the receiving antenna polarization vector $\mathbf{p}^0_{rec}$ allows expressing the component of electrical field intensity $\mathbf{E}_{rec}$ near the receiver, matched with the receiving antenna, in the form

$$\mathbf{E}_{rec\,match} = (\mathbf{p}^0_{rec})^{*T}\mathbf{E}_{rec} = (\mathbf{p}^0_{rec})^{*T}\mathbf{A}\mathbf{E}_{tg}/\sqrt{4\pi R^2}$$

In its turn, the transmitting antenna polarization vector $\mathbf{p}^0_{tr}$ allows expressing $\mathbf{E}_{tg} = C\mathbf{p}^0_{tr}/\sqrt{4\pi R^2}$, where $C = \text{const}$.

The value of RCS in the condition of the wave polarization transformation by the target can be given in the following form (see Section 1.3.4):

$$\sigma_{tg} = 4\pi R^2 \left|(\mathbf{p}^0_{rec})^{*T}\mathbf{E}_{rec}\right|^2 / \left|\mathbf{E}_{tg}\right|^2 = \left|(\mathbf{p}^0_{rec})^{*T}\mathbf{A}\mathbf{p}^0_{tr}\right|^2 \qquad (1.5)$$

Matched polarization reception or quadrature processing of the reception requires two orthogonal polarizations: $\sigma_{tg} = \left|\mathbf{A} \cdot \mathbf{p}^0_{tr}\right|^2$.

The own polarization basis of the target permits us to represent a polarization scattering matrix $\mathbf{A}$ through the diagonal matrix $\mathbf{M} = \mathbf{diag}(\mu_1, \mu_2)$ or

$$\mathbf{M} = \begin{bmatrix} \mu_1 & 0 \\ 0 & \mu_2 \end{bmatrix} \qquad (1.6)$$

It has the diagonal elements in the form of $\sqrt{\sigma_{1M}}e^{j\arg\mu_1} = \mu_1$ and $\sqrt{\sigma_{2M}}e^{j\arg\mu_2} = \mu_2$, which are the eigenvalues of matrices $\mathbf{A}$ and $\mathbf{M} = \mathbf{diag}(\mu_1, \mu_2)$, where σ_{1M} and σ_{2M} are the maximum and minimum possible values of the target RCS. Then,

$$\mathbf{A} = \mathbf{U}^{*T}\mathbf{M}\mathbf{U},$$

where matrices $\mathbf{U}$ are the orthogonal unitary complex matrices $\mathbf{U}^{*T}\mathbf{U} = \mathbf{I}$. This operation is a standard operation of matrix $\mathbf{A}$ diagonalization. A special case of matrix $\mathbf{U}$ is the matrix of the polarization basis rotation $\mathbf{U} = \begin{Vmatrix} \cos\varphi & -\sin\varphi \\ \sin\varphi & \cos\varphi \end{Vmatrix}$, where φ is the rotation angle.

1.1.2 Doppler Transform for Signals of Arbitrary Bandwidth-Duration Product

Let us consider the movement of a point target, flying away from a monostatic radar with constant radial velocity v_r [5, 6]. The solid line in Figure 1.1 depicts uniform target motion, its range at time t_0' is denoted by r_0'. Dotted lines in Figure 1.1 show schematically propagation with the constant velocity of light c of a wave fraction, transmitted between time moments t_0' and t' and received between the time moments t_0 and t, so that target illumination occurs between the time moments $(t_0' + t_0)/2$ and $(t' + t)/2$.

Target ranges at these moments are $r_0 + v_r(t_0' + t_0)/2$ and $r_0 + v_r(t' + t)/2$. They determine the echo signal delays $t_0 - t_0' = 2[r_0 + v_r(t_0' + t_0)/2]/c$ and $t - t' = 2[r_0 + v_r(t' + t)/2]/c$. The difference of these delays is $\Delta t' - \Delta t = -\dfrac{v_r}{c}(\Delta t' + \Delta t)$. Here, $\Delta t' = t' - t_0'$ is the duration of the transmitted signal's fraction, and $\Delta t = t - t_0$ is the corresponding duration of that of the received signal. The time scale transformation law is

$$\Delta t' = \frac{1 - v_r/c}{1 + v_r/c}\Delta t \text{ or } \Delta t' \approx \left(1 - \frac{2v_r}{c} + \frac{2v_r^2}{c^2}\right)\Delta t \text{ for } \frac{2v_r}{c} \ll 1 \quad (1.7)$$

The latter means that, with the usual neglect of the quadratic term, a doppler frequency shift $F_D = 2v_r f_0/c = 2v_r/\lambda_0$ takes place, negative for

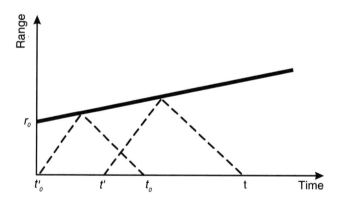

Figure 1.1 Clarification of signal's doppler transform.

$v_r > 0$ and positive for $v_r < 0$. Together with the doppler frequency shift, the signal's envelope is stretched for $v_r > 0$ or compressed for $v_r < 0$ also. This result corresponds to the special relativity theory, dealing with uniform movement of physical objects.

One can use (1.7) also for short signal fractions in the case of nonuniform target movement $v_r = v_r(t)$. For start time $t'_0 = 0$ of illumination and arbitrary signal duration

$$t' = t'(t) = t_0 + \int_{t_0}^{t} \frac{1 - v_r(s)/c}{1 + v_r(s)/c} ds \approx t - \frac{2}{c} r_0 - \frac{2}{c} \int_{t_0}^{t} v_r(s) ds = t - \frac{2r(t)}{c}$$

(1.8)

Neglect of quadric term in (1.7) leads to the errors in doppler frequency $\delta F_D = (2v_r^2/c^2) f_0$ and in phase $\delta\phi = 360° T \delta F_D$, where T is signal duration. These errors for $v_r = 1000$ m/s, $f_0 = 3 \cdot 10^{10}$ Hz, and $T = 0.3$ s are $\delta F_D = 0.7$ Hz, $\delta\phi = 24°$, so they can usually be neglected.

Hence, the doppler effect can be considered approximately as a result of distance variations, observed in the final reception moments. For time-extended real signals, it must be estimated not only as a frequency change, but as a stretching or compression of the signal envelope too.

1.1.2.1 Doppler Transform Example for an Extended Wideband Signal

Let us consider for the illumination Gaussian chirped pulse

$$U(t) = \exp\{-\pi[(1/\tau_p^2) + jK]t^2 + j2\pi ft\},$$

where τ_p is the signal duration at the level of $e^{-\pi/4} \approx 0.46$, f is the carrier frequency, Δf is the frequency deviation, and $K = \Delta f/\tau_p$ is the ratio of the frequency deviation to pulse duration. The pulse scattered by a point target is also Gaussian and chirped, but its parameters are changed; so parameters f and Δf are multiplied by $(1 - 2v_r/c)$ and parameter τ_p is divided by $(1 - 2v_r/c)$. Parameter K is multiplied by $(1 - 2v_r/c)^2 \approx 1 - 4v_r/c$, thus it receives the increment $\Delta K \approx -4(v_r/c)K$.

Matched processing presumes accounting for all the mentioned factors. Such a processing and range-velocity ambiguity function will be considered for stepped-frequency signals in Section 2.4.2.

1.2 Analog Methods of Scattering Simulation

The following analog methods are used to simulate backscattering in real target flights: full-scale experiments, scaled electrodynamic simulation, and scaled hydroacoustic simulation.

Full-scale experiments include the method of dynamic measurements and the method of statistical measurements [2, 7, 8]. Dynamic characteristics are obtained in the process of real flights using standard or instrumentation radar. The characteristics determined include: (1) the values of target radar cross sections and polarization scattering matrix elements at some fixed frequencies, (2) target's echoes to broadband illumination pulses at different carrier frequencies and to very short video pulses, and (3) modulation, fluctuation, and other statistical target's characteristics. Full-scale experiments are expensive. They are usually carried out only for the RCS estimation and face difficulties in case of the evaluation of recognition characteristics (mentioned in the Preface) for various illumination signals.

Scaled electrodynamic simulation is carried out by means of testing devices similar to those used in full-scale static simulation or in the anechoic chamber [2, 7, 8]. Great attention is paid to the plane wavefront forming near the target in anechoic chambers by means of special collimators, in particular. Characteristics of a real conductive target (index "tg") can be reproduced by the characteristics of conductive scaled models (index "md") if the likeness conditions are met:

$$\frac{l_{tg}}{l_{md}} = \frac{\lambda_{tg}}{\lambda_{md}} = \sqrt{\frac{\sigma_{tg}}{\sigma_{md}}} = \frac{t_{tg}}{t_{md}} \qquad (1.9)$$

These conditions connect the target's parameters with those of its model: (1) linear dimensions l_{tg}, l_{md}, (2) wavelengths λ_{tg}, λ_{md}, (3) radar cross sections σ_{tg}, σ_{md}, and (4) time duration of target and model response t_{tg}, t_{md}. The first ratio characterizes the required model size; the second one, wavelength; the third one allows us to recalculate the model RCS (diagonal elements of polarization scattering matrix) into those of real target. The last ratio allows us to evaluate the impulse responses of the target. Instead of wideband signal generation, small pulse-by-pulse frequency agility is sometimes used (see Section 2.4). Backscattered signals are subjected to phase detection using reflection from an external small-sized standard scatterer as a reference signal. The results are digitized and subjected to the fast Fourier transform (FFT). Multifrequency measurement of the polarization scattering matrix elements is carried out on the basis of models [2]. Scaled electrodynamic simulation is used for some recognition characteristic evaluation [7–11], but only for limited target numbers and usually without the motion consideration of them and their parts.

Scaled hydroacoustic simulation is based on the similarity of electromagnetic and acoustic wave propagation in isotropic media [12]. This similarity does not include polarization, which is the feature of electromagnetic waves that is absent in acoustics. An advantageous feature of hydroacoustic simulation is the significant decrease of acoustic wave propagation velocity v relative to the light velocity, which sharply reduces the wavelength, frequency bandwidth, model's dimensions, and dimensions of propagation tract and antennas:

$$\frac{l_{tg}}{l_{md}} = \frac{\lambda_{tg}}{\lambda_{md}} = \sqrt{\frac{\sigma_{tg}}{\sigma_{md}}} = \frac{ct_{tg}}{vt_{md}} = \frac{cf_{md}}{vf_{tg}} = \frac{cB_{md}}{vB_{tg}} \qquad (1.10)$$

Here, f_{tg}, f_{md} are the carrier frequencies; B_{tg}, B_{md} are the frequency bandwidths; c is the velocity of light; and v is the acoustic wave velocity. Examples of hydroacoustic simulation and target's signature determination were given for computerized water pool of 1m × 0.5m × 2m in [12]. Such simulation allows using different illumination signals, taking into account the motion of targets. The failures of this method are the lack of electromagnetic wave polarization simulation and the large attenuation of acoustic waves in water compared to electromagnetic wave attenuation in the atmosphere.

1.3 Computer Methods of Scattering Simulation

Computer simulation is based on approximate solutions of scattering-diffraction problems. These problems are connected with the choice of description method for the targets' surface. For centimeter and short decimeter waves, when the target surface can be described by a set of simpler ones (Section 1.3.1), the complex scattering problem is simplified. This method is detailed below (Sections 1.3.2–1.3.6) using coordinate transformations (Sections 1.3.2–1.3.3) and simplest bodies backscattering data (Section 1.3.6). Lastly, we discuss qualitatively the application limits of the chosen simplest component simulation method (Section 1.3.7).

1.3.1 Simplest Component and Other Methods of Target Surface Description and Calculation of Scattering

For the scattering calculation in centimeter and short decimeter radio wave bands (K, K_u, X, S, C, L) the quasioptical simplest component method

[2–4, 6, 12, 13] will be widely used (Chapters 1–6). This method reduces computational expenses while providing acceptable calculation accuracy. The target's airframe, wings, engine's pod, tail group, and outboard equipment are described with a wide set of simple bodies: quadric surfaces, plates, wedges, thin wires, disks, etc., for which sufficiently precise approximate theoretical relations have been already found. Available experimental results are also used for the cockpits, antenna modules, air intakes, engine's nozzles, etc. [3, 4]. The shadowing (masking) and rescattering effects are considered in analytical equations. The target is considered then as a simple multielement secondary radiator. The rescattering and multiple rescattering contributions in radar echo are increased with an increase of wavelength.

On parity with the simplest component method, other quasioptical methods (methods of physical optics) can be used. In Chapter 7 we will discuss the quasioptical methods without introducing the simplest components, in particular, the facet method that considers a target's surface airframe as consisting of facets (patches) and the combination of quasioptical methods with strict solutions [13].

As was pointed out, the simplest components and other quasioptical methods are restricted in wave band. For waves significantly shorter than K band, the target's surface cannot always be considered as smooth, so diffuse scattering has to be taken into account. For waves significantly longer than L band, multiple rescattering and associated resonance effects require the use of other methods. The wire method is often used where the target is considered as a set of thin wires [14]. The exact method of integral equations can be realized numerically on a computer for various objects [13]. Due to growth of computational expense, however, this method is not applicable yet in the high frequency domain.

1.3.2 Coordinate Systems and Coordinate Transforms Neglecting Earth's Curvature

Various coordinate systems must be introduced to simulate the signal backscattered from a moving target with moving elements by using the simplest component electrodynamic method:

- Radar's Cartesian coordinate system and the analogous spherical one—in the case of bistatic radar let us agree that they are referred to its transmitting part;

- Target's Cartesian coordinate system tied to target body and used in electrodynamic calculations (target body coordinate system);

- Cartesian coordinate systems of the simplest elements $\nu = 1, 2, \ldots,$ N_Σ (local coordinate systems).

Aerial target location in three-dimensional (3D) space is determined by six parameters (Figure 1.2). Three parameters X_{tg}, Y_{tg}, Z_{tg} determine the target center of mass location in the radar's Cartesian coordinate system $O_{rad}XYZ$. The OX and OZ axes lie in the Earth surface plane; the OY axis is directed upwards. Vector $\mathbf{R}_{tg} = \| X_{tg} \quad Y_{tg} \quad Z_{tg} \|^T$ also describes the target's mass center location.

Three rest parameters, namely course angle ψ, pitch angle θ and roll angle γ, determine the Cartesian target body coordinate system's $O_{tg}\xi\eta\zeta$ orientation. The $O_{tg}\xi$ axis of this system is directed along the target longitudinal axis to the nose, the $O_{tg}\eta$ axis is directed upwards, and the $O_{tg}\zeta$ axis is directed along the target right wing, so that the axes $O_{tg}\xi$, $O_{tg}\eta$, $O_{tg}\zeta$ make up the right three. Course angle ψ is the angle between the projection of the target longitudinal axis onto the horizontal plane $O_{rad}XZ$ and $O_{rad}X$ axis. Pitch angle θ is the angle between the target longitudinal axis and the horizontal plane. Roll angle γ is the angle between the target transversal axis $O_{tg}\zeta$ and the horizontal plane. Rotation direction is assumed to be positive if it is seen counterclockwise for the observer at the end of the rotation axis. In the case of neglecting Earth's curvature, the horizontal plane introduced here is the horizontal plane $O_{rad}XZ$ of radar coordinate system. In Section 1.3.3 the spherical Earth case will be considered. Then, the horizontal planes underlying the target will be taken into consideration.

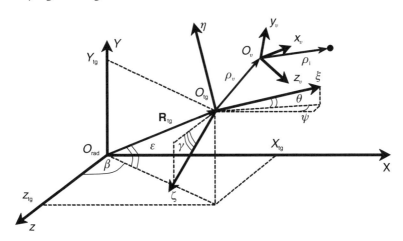

Figure 1.2 Coordinate system of radar $O_{rad}XYZ$, target $O_{tg}\xi\eta\zeta$, and local $O_\nu x_\nu y_\nu z_\nu$, used in backscattering simulation neglecting the Earth's curvature.

Interrelation of coordinates in the Cartesian radar system $O_{rad}XYZ$ and in the corresponding spherical one $O_{rad}R\beta\epsilon$ is described by equations

$$R = \sqrt{X^2 + Y^2 + Z^2}, \quad \beta = \operatorname{atan}\frac{Z}{X}, \quad \epsilon = \operatorname{atan}\frac{Y}{\sqrt{X^2 + Z^2}} \quad (1.11)$$

$$X = R\cos\epsilon\cos\beta, \quad Y = R\sin\epsilon, \quad Z = R\cos\epsilon\sin\beta \quad (1.12)$$

Recalculation of coordinates from the target coordinate system into the radar one can be made [15] by means of matrix function $\mathbf{H} = \mathbf{H}(\psi,\ \theta,\ \gamma)$ depending on rotation angles $\psi,\ \theta,\ \gamma$ (rotation matrix):

$$\left\| X \quad Y \quad Z \right\|^T = \left\| X_{tg} \quad Y_{tg} \quad Z_{tg} \right\|^T + \mathbf{H}^{-1}\left\| \xi \quad \eta \quad \zeta \right\|^T, \quad (1.13)$$

$$\left\| \xi \quad \eta \quad \zeta \right\|^T = \mathbf{H}^T\left\| X - X_{tg} \quad Y - Y_{tg} \quad Z - Z_{tg} \right\|^T \quad (1.14)$$

The rotation matrix for complex $\psi,\ \theta,\ \gamma$ rotation is the product of three simpler rotation matrices for separate rotations on each of these angles $\mathbf{H}(\psi,\ \theta,\ \gamma) = \mathbf{H}(\psi)\mathbf{H}(\theta)\mathbf{H}(\gamma)$, where (see Figure 1.3)

$$\mathbf{H}(\psi) = \left\|\begin{matrix} \cos\psi & 0 & \sin\psi \\ 0 & 1 & 0 \\ -\sin\psi & 0 & \cos\psi \end{matrix}\right\|, \quad \mathbf{H}(\theta) = \left\|\begin{matrix} \cos\theta & -\sin\theta & 0 \\ \sin\theta & \cos\theta & 0 \\ 0 & 0 & 1 \end{matrix}\right\|,$$

$$\mathbf{H}(\gamma) = \left\|\begin{matrix} 1 & 0 & 0 \\ 0 & \cos\gamma & -\sin\gamma \\ 0 & \sin\gamma & \cos\gamma \end{matrix}\right\|$$

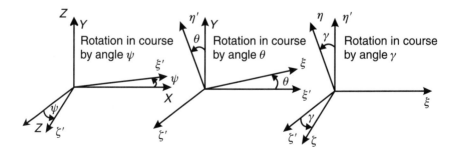

Figure 1.3 Clarification of coordinate system rotations in course, pitch, and roll angle.

Alternatively, the rotation matrix $\mathbf{H}(\psi, \theta, \gamma)$ is

$$\mathbf{H}(\psi, \theta, \gamma) = \tag{1.15}$$

$$\begin{Vmatrix} \cos\psi\cos\theta & -\cos\psi\sin\theta\cos\gamma + \sin\psi\sin\gamma & \cos\psi\sin\theta\sin\gamma + \sin\psi\cos\gamma \\ \sin\theta & \cos\theta\cos\gamma & -\cos\theta\sin\gamma \\ -\sin\psi\cos\theta & \cos\psi\sin\gamma + \sin\psi\sin\theta\cos\gamma & \cos\psi\cos\gamma - \sin\psi\sin\theta\sin\gamma \end{Vmatrix}$$

It is orthogonal; therefore, $\mathbf{H}^{\mathrm{T}} = \mathbf{H}^{-1}$.

Recalculation of unit vector $\mathbf{R}^0$ of incident wave propagation direction from the radar coordinate system into the target one is also made by the use of the matrix $\mathbf{H}$ transposed in this case:

$$\begin{Vmatrix} R_\xi^0 & R_\eta^0 & R_\zeta^0 \end{Vmatrix}^{\mathrm{T}} = \mathbf{H}^{\mathrm{T}}(\psi, \theta, \gamma)\begin{Vmatrix} R_X^0 & R_Y^0 & R_Z^0 \end{Vmatrix}^{\mathrm{T}} \tag{1.16}$$

Components of the unit vector $\mathbf{R}^0$ entered into (1.16) can be presented as coordinates (1.12) of the end of this unit vector, beginning at the coordinates' origin:

$$\begin{Vmatrix} R_X^0 & R_Y^0 & R_Z^0 \end{Vmatrix}^{\mathrm{T}} = \begin{Vmatrix} \cos\epsilon\cos\beta & \sin\epsilon & \cos\epsilon\sin\beta \end{Vmatrix}^{\mathrm{T}} \tag{1.17}$$

Local coordinate systems $O_\nu x_\nu y_\nu z_\nu$, $\nu = 0, 1, 2, \ldots, N_\Sigma - 1$ have their origins O_ν in the points ξ_ν, η_ν, ζ_ν of the target body coordinate system $O_{\mathrm{tg}}\xi\eta\zeta$, described by vectors $\boldsymbol{\rho}_\nu = \begin{Vmatrix} \xi_\nu & \eta_\nu & \zeta_\nu \end{Vmatrix}$, and their orientations are described by angles ψ_ν, θ_ν, γ_ν. Unit vector recalculation, analogous to (1.16), can be provided using values $\mathbf{H}(\psi_\nu, \theta_\nu, \gamma_\nu)$ of rotation matrix function (1.15):

$$\begin{Vmatrix} R_\xi^0 & R_\eta^0 & R_\zeta^0 \end{Vmatrix}^{\mathrm{T}} = \mathbf{H}(\psi_\nu, \theta_\nu, \gamma_\nu)\begin{Vmatrix} R_{x\nu}^0 & R_{y\nu}^0 & R_{z\nu}^0 \end{Vmatrix}^{\mathrm{T}} \tag{1.18}$$

The Aspect Angles

For the convenience of the program's use and the follow-up consideration, we can introduce the course-, pitch-, and roll-aspect angles of a target using the unit target radius-vector $\mathbf{R}^0$ and the coordinate unit vectors $\mathbf{Y}^0$, $\boldsymbol{\xi}^0$, and $\boldsymbol{\zeta}^0$. The course-aspect angle a_ψ from the tail of a target and a_ψ from its nose are defined as the angle between the projections of unit vectors $\pm\boldsymbol{\xi}^0$ and $\mathbf{R}^0$ on the $O_{\mathrm{rad}}XZ$ plane of the $O_{\mathrm{rad}}XYZ$ coordinate system.

$$a_\psi = 180° + a'_\psi \mathrm{acos}\{[\boldsymbol{\xi}^0 - \mathbf{Y}^0(\boldsymbol{\xi}^{0\mathrm{T}}\mathbf{Y}^0)]^\mathrm{T}$$
$$\cdot \, [\mathbf{R}^0 - \mathbf{Y}^0(\mathbf{R}^{0\mathrm{T}}\mathbf{Y}^0)] / \left|\boldsymbol{\xi}^0 - \mathbf{Y}^0(\boldsymbol{\xi}^{0\mathrm{T}}\mathbf{Y}^0)\right| \cdot \left|\mathbf{R}^0 - \mathbf{Y}^0(\mathbf{R}^{0\mathrm{T}}\mathbf{Y}^0)\right|\}$$

The pitch-aspect angle a_θ is defined as the angle between the projection of the unit vector $\boldsymbol{\xi}^0$ on the plane that passes through the unit vectors $\mathbf{R}^0$ and $\mathbf{Y}^0$, on the one hand, and the unit vector $\mathbf{R}^0$, on the other hand.

Designating by $\mathbf{S}^0 = \mathbf{R}^0 \times \mathbf{Y}^0 / |\mathbf{R}^0 \times \mathbf{Y}^0|$ the unit vector of the cross product $\mathbf{R}^0 \times \mathbf{Y}^0$, we have

$$a_\theta = \mathrm{acos}\{[\boldsymbol{\xi}^0 - \mathbf{S}^0(\boldsymbol{\xi}^{0\mathrm{T}}\mathbf{S}^0)]^\mathrm{T}\mathbf{R}^0 / \left|\boldsymbol{\xi}^0 - \mathbf{S}^0(\boldsymbol{\xi}^{0\mathrm{T}}\mathbf{S}^0)\right|\}$$

The roll-aspect angle a_γ is defined as the angle between the projection of the unit vector $\boldsymbol{\zeta}^0$ on the plane that passes through the unit vectors $\mathbf{R}^0$ and $\mathbf{Y}^0$, on the one hand, and the unit vector $\mathbf{R}^0$, on the other hand,

$$a_\gamma = \mathrm{acos}\{[\boldsymbol{\varsigma}^0 - \mathbf{S}^0(\boldsymbol{\varsigma}^{0\mathrm{T}}\mathbf{S}^0)]^\mathrm{T}\mathbf{R}^0 / \left|\boldsymbol{\varsigma}^0 - \mathbf{S}^0(\boldsymbol{\varsigma}^{0\mathrm{T}}\mathbf{S}^0)\right|\}$$

All three aspect angles are calculated approximately by our program of the flight simulations. In the follow-up parts of the book we shall mainly use only the course-aspect angle assuming zero pitch and roll angles, hereinafter called the aspect angle.

1.3.3 Coordinate Systems and Coordinate Transforms Accounting for Earth's Curvature

The target orientation angles θ, γ were measured relative to a horizontal plane common to radar and target. As the target's distance grows, the horizontal plane tangent to the spherical surface underlying the target rotates relative to the horizontal plane at the radar. To employ the results of Section 1.3.2, let us introduce an auxiliary coordinate system $O_\mathrm{rad}X'Y'Z'$ with origin in the point O_rad of the radar position. It corresponds to the rotation of the coordinate system $O_\mathrm{rad}XYZ$ in the $O_\mathrm{rad}O_\mathrm{tg}O_\mathrm{E}$ "radar-target-Earth's center" plane. The axis $O_\mathrm{rad}Y$ is converted into the $O_\mathrm{rad}Y'$ one parallel to the $O_\mathrm{E}O_\mathrm{tg}$ line that passes through the Earth's center O_E and the target coordinate system origin O_tg [Figure 1.4(a)]. As it was before, the orientation angles θ, γ can be measured relative to the horizontal plane $O_\mathrm{rad}X'Z'$ of the auxiliary coordinate system $O_\mathrm{rad}X'Y'Z'$. All this allows us to employ the results of Section 1.3.2. An additional problem, however, arises when

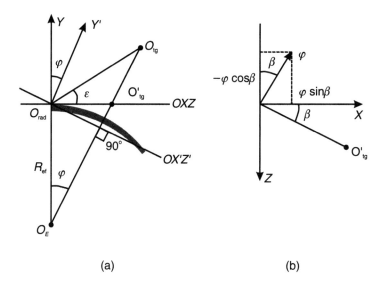

(a)   (b)

Figure 1.4 (a) The "radar-target-Earth's center" $O_{rad}O_{tg}O_E$ plane, coordinate axes $O_{rad}Y$, $O_{rad}Y'$ lying in this plane and traces $O_{rad}XZ$, $O_{rad}X'Z'$ of coordinate axes' intersection with $O_{rad}O_{tg}O_E$ plane. (b) The radar horizon plane $O_{rad}XZ$, projection O'_{tg} of target coordinate system origin onto it and vector φ of coordinate system $O_{rad}XYZ$ rotation into $O_{rad}X'Y'Z'$.

converting data from $O_{rad}X'Y'Z'$ into $O_{rad}XYZ$ coordinate system, and inversely.

For a solution to this problem, let us introduce the rotation vector φ [Figure 1.4(b)] with the absolute value of the "radar-target" geocentric angle [Figure 1.4(a)]. Vector φ is oriented perpendicularly to the rotation plane, so that rotation by angle φ is seen counterclockwise for an observer at its end. For given target azimuth β, this vector has components 0, $-\varphi\cos\beta$, and $\varphi\sin\beta$ along the axes $O_{rad}Y$, $O_{rad}Z$, and $O_{rad}X$.

Repeating the considerations connected with Figure 1.3 and (1.13) through (1.15), one may obtain the recalculation matrix $\mathbf{H}(0, -\varphi\cos\beta, \varphi\sin\beta)$ from the main radar coordinate system $O_{rad}XYZ$ into the auxiliary one $O_{rad}X'Y'Z'$. The recalculation matrix from the main radar coordinate system into the target one is reduced to the matrix product $\mathbf{H}(0, -\varphi\cos\beta, \varphi\sin\beta)\mathbf{H}(\psi, \theta, \gamma)$. Equation (1.16) for the unit vector of incident vector recalculation into target coordinate system becomes

$$\left\| R_{\xi}^0 \quad R_{\eta}^0 \quad R_{\zeta}^0 \right\|^{\mathrm{T}} = \tag{1.19}$$

$$\mathbf{H}^{\mathrm{T}}(\psi, \, \theta, \, \gamma)\,\mathbf{H}^{\mathrm{T}}(0, \, -\varphi\cos\beta, \, \varphi\sin\beta)\left\| R_X^0 \quad R_Y^0 \quad R_Z^0 \right\|^{\mathrm{T}}$$

The geocentric angle $\varphi \approx -L/R_{ef}$ introduced above and entered into (1.19) is determined by the target range L along the curved Earth and effective Earth radius R_{ef}.

1.3.4 Peculiarities of the Simplest Component Method Employment

An airframe is described using its drawing, and its moving parts are described on the basis of their parameters (Section 1.5). The target surface is divided into $N_\Sigma = 60, \ldots, 200$ independent elementary surfaces $F(\mathbf{r}_\nu) = F_\nu(x_\nu, y_\nu, z_\nu) = 0$, $\nu = 0, 1, 2, \ldots, N_\Sigma - 1$, considered as conductive [2, 3, 8] and called *approximating surfaces of the first kind*. They are described with limited double curved surfaces, straight or bent wedges, cones, truncated cones, cylinders, tori, ogives, and plates in their own local coordinate systems $O_\nu x_\nu y_\nu z_\nu$ (Figure 1.5).

A point on the νth first-kind surface is presumed to belong to the target if it lies inside some auxiliary limiting surfaces $\Phi_{\nu k}(\mathbf{r}_\nu) < 0$, $k = 0, 1, 2, \ldots, K_\nu - 1$. Here, k is the number of a surface $\Phi_{\nu k}$ limiting the νth scattering surface, and K_ν is the overall number of such surfaces. Each approximating and limiting surface is defined originally in its local (canonical) coordinate system $O_\nu x_\nu y_\nu z_\nu$ with the equations $F_\nu(\mathbf{r}_\nu) = 0$ and $\Phi_{\nu k}(\mathbf{r}_\nu) = 0$. For example, part of a cylindrical surface (Figure 1.6) given by the equation $F_\nu(\mathbf{r}_\nu) = F_\nu(x_\nu, y_\nu, z_\nu) = x_\nu^2/a_\nu^2 + y_\nu^2/b_\nu^2 = 0$ can be limited

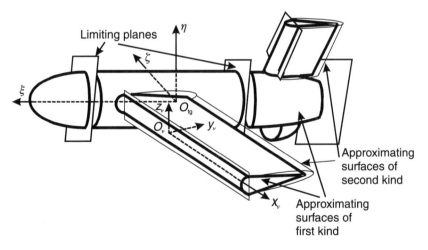

Figure 1.5 Clarification of target surface description with the simplest components. The two kinds of approximating surfaces are shown here: the first kind used to calculate the backscattering and to consider shadowing (masking); the second kind used to consider shading only.

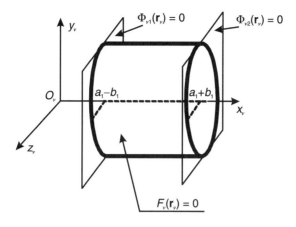

Figure 1.6 Example of the cylindrical surface limited with two planes.

by a pair $K_\nu = 2$ of planes $\Phi_{\nu k}(\mathbf{r}) = x_\nu - c_\nu \pm d_\nu = 0$, $k = 1, 2$ normal to its axis. The pair of planes can be described by the equation of an elliptical cylinder $\Phi_{\nu k}(\mathbf{r}_\nu) = (x_\nu - c_\nu)^2 - d_\nu^2 = 0$ having the second infinite axis. Parameters a_ν, b_ν correspond here to the approximating surface, and parameters c_ν, d_ν to the limiting ones. Such a method of introducing the limiting surfaces was widely used in the design of simulation programs.

Along with the natural problem of target element's RCS representation, the problems of shadowing of some of their elements with others have to be solved as well.

For the solution of the first problem, the target's fuselage is approximated by two paraboloids and a cylinder. Wing edges and those of the tail group are approximated by wedges; their parameters are the length and the external angle in radians. Fractures at the quadric surface joints are approximated with wedges by curved ribs. The external aperture and radius of curvature must be specified in this case. The edges of the air intakes are approximated by torus parts or thin wedges. The engine compressor and turbine blade, and propeller blade approximations are considered below (Section 1.5). Onboard antennas can be described by their parameters: coordinates of their centers, unit vectors normal to the apertures, operational wavelengths, and focal distances (for reflector antennas).

For the solution of shadowing problems Yu. V. Sopelnik introduced *approximating surfaces of the second kind* $\Psi_\mu(\mathbf{r}_\nu) = 0$, $\mu = 0, 1, 2, \ldots,$ $M_\mu - 1$. They must have a quadric form and be adjacent to some sharp elements, embracing them (Figure 1.5). Their quadric form makes it easier to consider the shadowing (see Section 1.3.6). Quadric approximating surfaces of the first kind can be used as approximating surfaces of the second kind.

One or several bright points, bright lines, or bright areas (see Section 1.3.6) can substitute for each νth scattering element. The number of accounted bright elements $N \neq N_\Sigma$ varies due to the shadowing and specular reflection effects. For the plane surfaces one must be careful in neglecting the shadowed bright points (see Section 1.5.3).

1.3.5 General Equations of Scattering for the Far-Field Zone and Arbitrary Signal Bandwidth-Duration Product

Angular resolution of target elements is not assumed. We consider the following:

1. Coordinate and orientation changes of the target and target elements, accompanied by the doppler transform (1.8), when neglecting Earth's curvature;

2. The polarization transform (1.3) through (1.5);

3. Matched processing (filtering) of partial reflections;

4. Superposition of reflections $X(t)$ at the output of the linear part of the receiver;

5. System linearity, justifying the change of the order of calculation of the matched processing and reflections' superposition consideration described in (3) and (4) above;

6. Possible covering of some bright elements with radar absorbing material (RAM);

7. Possible spacing $\mathbf{L}$ of the receiving antenna from the transmitting one as in the case of bistatic radar. In case of the simplest monostatic radar, mainly considered in the book, the value $\mathbf{L} = 0$ will be used. In case of more general monostatic radar with a separate receiving antenna small values of $\mathbf{L}$ other than zero will be taken into consideration, for instance in the glint simulation.

We do not discuss here ultra-wideband signals and Earth-surface influence. Modification of computations in the ultra-wideband case will be considered in Chapter 2. Low-altitude aerial targets will be considered in Chapter 5.

General Equation

The factors considered lead to the following general equation:

$$\dot{E}(t, \mathbf{L}) = (\mathbf{p}_{\text{rec}}^0)^{*\text{T}} \left[\sum_{i=1}^{N} \mathbf{A}_i(\mathbf{R}^0, \mathbf{L}) U(t - \Delta t_i) e^{-j 2 \pi f \Delta t_i} 10^{-Q_{\text{Ab}i}/20} \right] \mathbf{p}_{\text{tr}}^0$$

(1.20)

Here,

$\mathbf{A}_i(\mathbf{R}^0, \mathbf{L})$ is the polarization scattering matrix (1.3) of the ith bright element in the radar basis. For small values $\mathbf{L}$ it will be assumed that $\mathbf{A}_i(\mathbf{R}^0, \mathbf{L}) \approx \mathbf{A}_i(\mathbf{R}^0)$;

$\mathbf{p}_{\text{rec}}^0, \mathbf{p}_{\text{tr}}^0$ are the polarization vectors (1.4) of the receiving and transmitting antennas;

$U(t)$ is the value of complex envelope of the matched filter's output at a time t;

$\Delta t_i = [\,|\mathbf{R}_i(t)| + |\mathbf{R}_i(t) - \mathbf{L}|\,]/c$ is the delay corresponding to the ith bright element;

$\mathbf{R}_i(t) = \mathbf{R}(t) + \boldsymbol{\rho}_i(t)$ is the radius-vector of the ith bright element in the radar coordinate system;

$\mathbf{R}(t)$ is the radius-vector of the target coordinate system origin in the radar one;

$\boldsymbol{\rho}_i(t) = \boldsymbol{\rho}_\nu(t) + \mathbf{r}_i(t)$ is the radius-vector of the ith bright element in the target coordinate system;

$\boldsymbol{\rho}_\nu(t)$ is the radius-vector of the νth local coordinate system origin in the target one;

$\mathbf{r}_i(t)$ is the radius-vector of the ith bright element in the νth local coordinate system;

f is the carrier frequency;

c is the velocity of light in free space;

$Q_{\text{Ab}i}$ is the absorption coefficient in dB of the ith bright element's RAM covering for the given carrier frequency and signal bandwidth.

Linear Approximation of the General Equation. This accounts for small values of $\mathbf{L}$ in the case of a separate receiving antenna (Section 6.2) and takes the form

$$\dot{E}(t, \mathbf{L}) \approx \dot{E}(t, 0) + \mathbf{L}^{\text{T}} \frac{d\dot{E}(t, 0)}{d\mathbf{L}}, \text{ where } \frac{d\dot{E}(t, 0)}{d\mathbf{L}} = \frac{d\dot{E}(t, \mathbf{L})}{d\mathbf{L}} \bigg|_{\mathbf{L}=0}$$

(1.21)

Let us mention that the vector derivative of a scalar function $f(x, y, z) = f(\mathbf{r})$, where $\mathbf{r} = x\mathbf{x}^0 + y\mathbf{y}^0 + z\mathbf{z}^0$, is the gradient of this function $\frac{df}{d\mathbf{r}} = \frac{df}{dx}\mathbf{x}^0 + \frac{df}{dy}\mathbf{y}^0 + \frac{df}{dz}\mathbf{z}^0 = \mathrm{grad} f(x, y, z)$.

The Operational Form of the General Equation. It is convenient for its succeeding development under conditions of wave propagation above the underlying surface (Section 6.2). It is defined by a set of equations for $E(t, 0) = E(t)$:

$$E(t) = \int_0^\infty E(p)e^{pt}dp, \quad E(p) = (\mathbf{p}_{\mathrm{rec}}^0)^{*\mathrm{T}}\mathbf{A}(p)\mathbf{p}_{\mathrm{tr}}^0 U_\omega(p), \quad (1.22)$$

$$U_\omega(p) = \int_0^\infty [U(t)e^{j\omega t}]e^{-pt}dt$$

where $\mathbf{A}(p)$ is the operational form of the target polarization scattering matrix

$$\mathbf{A}(p) = \sum_{i=1}^N \mathbf{A}_i(\mathbf{R}^0)e^{-pt_i}10^{-Q_{\mathrm{Ab}i}/20} \quad (1.23)$$

1.3.6 Use of the Simplest Components' Initial Data

The exact calculation of the scattered electromagnetic field for some simplest bodies is carried out using solutions to Maxwell's equation. Exact solutions to scattering (diffraction) problems are known only for the simplest cases. Such problems are known as "model" (ellipsoid, sphere, cylinder, wedge, half-plane) problems and are solved by methods of mathematical physics. The solutions are used for:

- The evaluation of principal limitations to succeeding approximations;

- The construction of approximate methods of short-wave asymptotic calculation (geometric and physical optics, geometrical and physical diffraction theories) based partly on the "model" solutions.

It is not necessary to discuss these theories here in detail because of their excellent exposition given by E. F. Knott and others in [2]. Let us consider available data about the bright elements given in Tables 1.1 through 1.3 [2–4, 7, 8, 16].

Table 1.1
Expressions for the Specular Point and Line Coordinate Calculation of Smooth Surfaces

No.	Surface Type	Surface's Canonical Equation	"Bright" Point Coordinates*
1.	Ellipsoid	$\dfrac{x^2}{a^2} + \dfrac{y^2}{b^2} + \dfrac{z^2}{c^2} = 1$	$x = -\dfrac{R_x^0 a^2}{U};\ y = -\dfrac{R_y^0 b^2}{U};\ z = -\dfrac{R_z^0 c^2}{U};$ $U = \sqrt{(R_x^0)^2 a^2 + (R_y^0)^2 b^2 + (R_z^0)^2 c^2}$
2.	Elliptical paraboloid	$\dfrac{z^2}{p} + \dfrac{y^2}{q} = 2x$	$x = \dfrac{1}{2}\left(p\dfrac{(R_z^0)^2}{(R_x^0)^2} + q\dfrac{(R_y^0)^2}{(R_x^0)^2}\right);\ y = -q\dfrac{R_y^0}{R_x^0};\ z = -p\dfrac{R_z^0}{R_x^0}$
3.	Two-cavity hyperboloid	$\dfrac{z^2}{c^2} + \dfrac{y^2}{b^2} - \dfrac{x^2}{a^2} = -1$	$x = -\dfrac{R_x^0 a^2}{U};\ y = -\dfrac{R_y^0 b^2}{U};\ z = -\dfrac{R_z^0 c^2}{U};$ $U = \sqrt{(R_x^0)^2 a^2 + (R_y^0)^2 b^2 + (R_z^0)^2 c^2}$
4.	Elliptical cylinder	$\dfrac{z^2}{c^2} + \dfrac{y^2}{b^2} = 1$	$x_1 = 0;\ x_2 = 1;\ y_1 = y_2 = -\dfrac{R_y^0 b^2}{U};\ z_1 = z_2 = -\dfrac{R_z^0 c^2}{U};$ $U = \sqrt{(R_y^0)^2 b^2 + (R_z^0)^2 c^2}$
5.	Parabolic cylinder	$y^2 = 2pz$	$x_1 = 0;\ x_2 = 1;\ y_1 = y_2 = -p\dfrac{R_y^0}{R_z^0};\ z_1 = z_2 = \dfrac{p}{2}\left(\dfrac{R_y^0}{R_z^0}\right)^2;$

*For cylindrical and conical surfaces, the coordinates of two points are forecited that determine the limits of bright line.

Table 1.1 (continued)

No.	Surface Type	Surface's Canonical Equation	"Bright" Point Coordinates*
6.	Hyperbolic cylinder	$\dfrac{z^2}{c^2} - \dfrac{y^2}{b^2} = 1$	$x_1 = 0;\ x_2 = 1;\ y_1 = y_2 = -\dfrac{R_y^0 b^2}{U};\ z_1 = z_2 = \dfrac{R_z^0 c^2}{U};$ $U = \sqrt{(R_z^0)^2 c^2 - (R_y^0)^2 b^2}$
7.	Elliptical cone	$\dfrac{z^2}{c^2} + \dfrac{y^2}{b^2} - \dfrac{x^2}{a^2} = 0$	$x_1 = y_1 = z_1 = 0;\ x_2 = a;\ y_2 = -\dfrac{R_y^0 b^2}{U};\ z_2 = \dfrac{R_z^0 c^2}{U};$ $U = \sqrt{(R_y^0)^2 b^2 + (R_z^0)^2 c^2}$
8.	Hyperbolic paraboloid**	$\dfrac{z^2}{p} - \dfrac{y^2}{q} = 2x$	$x = \dfrac{1}{2}\left(p\dfrac{(R_z^0)^2}{(R_x^0)^2} - q\dfrac{(R_y^0)^2}{(R_x^0)^2}\right);\ y = -q\dfrac{R_y^0}{R_x^0};\ z = p\dfrac{R_z^0}{R_x^0}$

*For cylindrical and conical surfaces, the coordinates of two points are forecited that determine the limits of bright line.

**The applicability of a hyperbolic paraboloid model in simulation is hampered by the alterable sign of its surface curvature. The bright point spreads out onto the bright line in this case, and the use of equation (1.26) becomes incorrect. It would be more legitimate to divide the curved bright line onto a set of small straight lines belonging to short cylinders of corresponding curvature radii.

Table 1.2
Expressions of Curvature Radii and Other Parameters of Smooth Surfaces

No. 1	Surface Type 2	Canonical Equation 3	Parametric Equation 4	Inverse Parametric Equation 5	Expressions of the Curvature Radii 6
1.	Ellipsoid	$\dfrac{x^2}{a^2}+\dfrac{y^2}{b^2}+\dfrac{z^2}{c^2}=1$	$x=a\sin u\cos v$; $y=b\sin v\sin u$; $z=c\cos v$	$u=\operatorname{atan}\left(\dfrac{ay}{bx}\right)$; $v=\operatorname{acos}\left(\dfrac{z}{c}\right)$	$R_1=1/k_1$; $R_2=1/k_2$; $k_1=H+\sqrt{H^2-K}$; $k_2=H-\sqrt{H^2-K}$; $H=\dfrac{1}{2}\dfrac{l\cdot G+nE-2mF}{(EG-F^2)^{3/2}}$; $K=\dfrac{l\cdot n-m^2}{(EG-F^2)^2}$; $G=(a\cos u\cos v)^2+(b\cos v\sin u)^2+(c\sin v)^2$; $E=(a\sin u\sin v)^2+(b\sin v\cos u)^2$; $F=\cos u\sin u\cos v\sin v(b^2-a^2)$; $l=abc\sin^3 v$; $m=0$; $n=abc\sin v$;
2.	Elliptical paraboloid	$\dfrac{z^2}{p}+\dfrac{y^2}{q}=2x$	$x=v^2/2$; $y=\sqrt{q}\,v\sin u$; $z=\sqrt{p}\,v\cos u$	$u=\operatorname{atan}\left(\dfrac{y}{z}\sqrt{\dfrac{p}{q}}\right)$; $v=\sqrt{2x}$	$R_1=1/k_1$; $R_2=1/k_2$; $k_1=H+\sqrt{H^2-K}$; $k_2=H-\sqrt{H^2-K}$; $H=\dfrac{1}{2}\dfrac{l\cdot G+nE-2mF}{(EG-F^2)^{3/2}}$; $K=\dfrac{l\cdot n-m^2}{(EG-F^2)^2}$; $G=v^2+q\sin^2 u+p\cos^2 u$; $E=v^2(q\cos^2 u+p\sin^2 u)$; $F=v\cos u\sin u(q-p)$; $l=\sqrt{pqv^3}$; $m=0$; $n=-\sqrt{pqv}$

Table 1.2 (continued)

No. 1	Surface Type 2	Canonical Equation 3	Parametric Equation 4	Inverse Parametric Equation 5	Expressions of the Curvature Radii 6
3.	Two-cavity hyperboloid	$\dfrac{z^2}{c^2} + \dfrac{y^2}{b^2} - \dfrac{x^2}{a^2} = -1$	$x = a\cosh v$; $y = b\sinh v \sin u$; $z = c\sinh v \cos u$	$u = \operatorname{atan}\left(\dfrac{cy}{bz}\right)$; $v = \ln(x/a + \sqrt{(x/a)^2 - 1})$	$R_1 = 1/k_1$; $R_2 = 1/k_2$; $k_1 = H + \sqrt{H^2 - K}$; $k_2 = H - \sqrt{H^2 - K}$; $H = \dfrac{1}{2}\dfrac{I \cdot G + nE - 2mF}{(EG - F^2)^{3/2}}$; $K = \dfrac{I \cdot n - m^2}{(EG - F^2)^2}$; $G = (a\sinh v)^2 + (b\cosh v \sin u)^2 + (c\cosh v \cos u)^2$; $E = (a\sinh v \cos u)^2 + (c\sinh v \sin u)^2$; $F = \cosh v \sinh v \cos u \sin u(b^2 - c^2)$; $I = abc\sinh^3 v$; $m = 0$; $n = abc\sinh v$
4.	Elliptical cylinder	$\dfrac{z^2}{c^2} + \dfrac{y^2}{b^2} = 1$	$y = b\sin t$; $z = c\cos t$		$R = \dfrac{b^2 c^2(y^2/b^4 + z^2/c^4)^{3/2}}{z^2/c^2 + y^2/b^2}$
5.	Parabolic cylinder	$y^2 = 2pz$	$y = t$; $z = \dfrac{t^2}{2p}$		$R = \dfrac{(y^2 + p^2)^{3/2}}{p^2}$

Table 1.2 (continued)

No. 1	Surface Type 2	Canonical Equation 3	Parametric Equation 4	Inverse Parametric Equation 5	Expressions of the Curvature Radii 6
6.	Hyperbolic cylinder	$\dfrac{z^2}{c^2} - \dfrac{y^2}{b^2} = 1$	$y = b\sinh t$; $z = a\cosh t$		$R = \dfrac{b^2 c^2 (y^2/b^4 + z^2/c^4)^{3/2}}{\left|z^2/c^2 - y^2/b^2\right|}$
7.	Elliptical cone	$\dfrac{z^2}{c^2} + \dfrac{y^2}{b^2} - \dfrac{x^2}{a^2} = 0$	$x = av$; $y = bv\sin u$; $z = cv\cos u$	$u = \operatorname{atan}\left(\dfrac{yc}{zb}\right)$; $v = \dfrac{x}{a}$	$R = \dfrac{b^2 c^2 + c^2 a^2 \sin^2 u + b^2 a^2 \cos^2 u)^{3/2} v}{(c^2 \cos^2 u + b^2 \sin^2 u + a^2)abc}$
8.	Hyperbolic paraboloid	$\dfrac{z^2}{p} - \dfrac{y^2}{q} = 2x$	$x = \dfrac{1}{2}v^2$; $y = \sqrt{p}\,v\cosh u$; $z = \sqrt{q}\,v\sinh u$	$u = \dfrac{1}{2}\ln\dfrac{1+b}{1-b}$, where $b = \dfrac{z}{y}\sqrt{\dfrac{p}{q}}$; $v = \sqrt{2x}$	$R_1 = 1/k_1$; $R_2 = 1/k_2$; $k_1 = H + \sqrt{H^2 - K}$; $k_2 = H - \sqrt{H^2 - K}$; $H = \dfrac{1}{2}\dfrac{l\cdot G + nE - 2mF}{(EG - F^2)^{3/2}}$; $K = \dfrac{l\cdot n - m^2}{(EG - F^2)^2}$; $G = v^2 + p\cosh^2 u + q\sinh^2 u$; $E = v^2(p\sinh^2 u + q\cosh^2 u)$; $F = v\cosh u \sinh u(p + q)$; $l = \sqrt{pq}\,v^3$; $m = 0$; $n = -\sqrt{pq}\,v$

Table 1.3
Geometric Parameters and RCSs of Ideally Conducting Surfaces' Bright Elements

No. 1	Reflector Type 2	Geometry Parameters 3	RCSs of Ideally Conducting Surfaces' "Bright" Elements 4	5
1.	Double curved surface [4, Ch. 4]		$$\sigma = \pi \left.\frac{\left(\frac{\partial F}{\partial x}\right)^2 + \left(\frac{\partial F}{\partial y}\right)^2 + \left(\frac{\partial F}{\partial z}\right)^2}{-Q}\right|_{P_0}$$ where $$Q = \begin{vmatrix} \dfrac{\partial^2 F}{\partial x^2} & \dfrac{1}{2}\dfrac{\partial^2 F}{\partial x\partial y} & \dfrac{1}{2}\dfrac{\partial^2 F}{\partial x\partial z} & \dfrac{\partial F}{\partial x} \\[2mm] \dfrac{1}{2}\dfrac{\partial^2 F}{\partial x\partial y} & \dfrac{\partial^2 F}{\partial y^2} & \dfrac{1}{2}\dfrac{\partial^2 F}{\partial y\partial z} & \dfrac{\partial F}{\partial y} \\[2mm] \dfrac{1}{2}\dfrac{\partial^2 F}{\partial x\partial z} & \dfrac{1}{2}\dfrac{\partial^2 F}{\partial y\partial z} & \dfrac{\partial^2 F}{\partial z^2} & \dfrac{\partial F}{\partial z} \\[2mm] \dfrac{\partial F}{\partial x} & \dfrac{\partial F}{\partial y} & \dfrac{\partial F}{\partial z} & 0 \end{vmatrix}$$	$F(x, y, z) = 0$ – surface equation. $P_0 = P_0(x_0, y_0, z_0)$ – "bright" point
2.	Ogive [4, Ch. 4, eqs. 5–7]		$$\sigma(\theta) = \frac{\lambda^2\tan^4\alpha}{16\pi\cos^6\theta(1 - \tan^2\alpha \cdot \tan^2\theta)^3}$$ $$\sigma(\pi/2 - \alpha) = \frac{a^2}{4\pi\tan^2(\alpha/2)}$$ $$\sigma(\theta) = \pi R^2\left(1 - \frac{R-a}{R\sin\theta}\right)$$	in $0 \le \theta < \pi/2 - \alpha$. in $\theta = \pi/2 - \alpha$. in $\left|\pi/2 - \alpha\right| < \theta \le \pi/2$. Here and below λ is the wavelength.

Table 1.3 (continued)

No. 1	Reflector Type 2	Geometry Parameters 3	RCSs of Ideally Conducting Surfaces' "Bright" Elements 4	5
3.	Torus [4, Ch. 4, eqs. 39–41]		$$\sigma = \frac{8\pi^3 ba^2}{\lambda}$$ $$\sigma_1(\theta) = \pi\left(\frac{ba}{\sin\theta}+b^2\right);\ \sigma_2(\theta) = \pi\left(\frac{ba}{\sin\theta}-b^2\right)$$	in $\theta = 0$, in $\theta > 0$. If $0 \le |\cos\theta| \le b/2a$ then σ_2 is absent.
4.	Wedge with straight rib of L length		$$\sigma_{\|\perp} = \frac{L^2}{\pi}\left|\frac{\sin\pi/n}{n}\left[\left(\cos\frac{\pi}{n}-1\right)^{-1} \mp \left(\cos\frac{\pi}{n}-\cos\frac{2\theta}{n}^{-1}\right)\right]\right|.$$ where $n = \alpha/\pi$. If the direction of incidence differs from the normal to the rib by a small angle $\delta\beta$ (not shown), then $$\frac{\sigma_{nonnormal}}{\sigma_{normal}} = g^2 \approx \frac{\lambda^2}{8\pi^2 L^2(\delta\beta)^2}.$$	$\sigma_{\|}$ is the RCS for the case when vector **E** is parallel to the rib; $\sigma_{\perp}$ is the RCS for the case when vector **H** is parallel to the rib. This formula is true for the directions of diffraction lying on the Keller cone and being far from those of specular reflection from the wedge faces.

Table 1.3 (continued)

No. 1	Reflector Type 2	Geometry Parameters 3	RCSs of Ideally Conducting Surfaces' "Bright" Elements 4	5
5.	Wedge with curved rib		$$\sigma_{\parallel,\perp} = \left|\frac{a}{k\cos v}\left|\left(\cos\frac{\pi}{n}-1\right)^{-1} \mp \left(\cos\frac{\pi}{n}-\cos\frac{2\theta}{n}\right)^{-1}\right|\right|^{1,2}.$$ where $n = \alpha/\pi$, a is the curvature radius in the point of diffraction, v is the angle between the principle normal to the rib in the diffraction point and the direction of diffraction.	This formula is true for the directions of diffraction lying on the Keller cone and being far from those of specular reflection from the wedge faces.
6.	Thin cylinder (wire of length L and diameter d) [4, Ch. 4, eqs. 26 and 27]		$$\sigma_{1,2} = \frac{\lambda^2\tan^2\theta\cos^4\Phi}{16\pi\left[\left(\frac{\pi}{2}\right)^2 + \ln^2\left(\frac{\lambda}{1.78\pi d\sin\theta}\right)\right]}, \theta \neq 90°,$$ $$\sigma_\Sigma = \frac{\pi L^2\cos^4\Phi}{\left(\frac{\pi}{2}\right)^2 + \ln^2\left(\frac{\lambda}{1.78\pi d}\right)}$$ in normal incidence to wire's axis. Both equations are given for $L > (2, \ldots, 3)\lambda$; $d \leq (1/10, \ldots, 1/8)\lambda$.	Two "brilliant" points shifted by $\lambda/(8\sin\theta)$ from the ends of the wire; Φ is the angle between the cylinder axis and vector $\mathbf{E}$ in linear polarization

Table 1.3 (continued)

No. 1	Reflector Type 2	Geometry Parameters 3	RCSs of Ideally Conducting Surfaces' "Bright" Elements 4	5
7.	Cylindrical surface. Disk is accounted for separately.		$\sigma_{\parallel \mid \perp} = \dfrac{a_i}{k\sin\theta}\left[\dfrac{\sin \pi/n}{n}\left(\left(\cos\dfrac{\pi}{n}-1\right)\right.\right.$ $\left.\left.\mp \left(\cos\dfrac{\pi}{n}-\cos\dfrac{2(\theta+\theta_i)}{n}\right)^{-1}\right)\right]^{-2}$. Here, $n = 3/2$; a_i is the curvature radius in the point S_1 (S_2); θ_i is the angle between the wedge reference face and "brilliant" line. For $\theta = \pi/2$ $\sigma = kaL^2$. In normal incidence to the generatrix.	Angle θ_i takes the value 0 or $\pi/2$ in this case Here and below, $k = 2\pi/\lambda$ is the wave number
8.	Truncated cone surface [3, Ch. 6, eqs. 6.3–61]. Disk is accounted for separately.		$\sigma_{\parallel \mid \perp} = \dfrac{a_i}{k\sin\theta}\left[\dfrac{\sin \pi/n}{n}\left(\left(\cos\dfrac{\pi}{n}-1\right)\right.\right.$ $\left.\left.\mp \left(\cos\dfrac{\pi}{n}-\cos\dfrac{2(\theta-\theta_i)}{n}\right)^{-1}\right)\right]$. Here, $n = 3/2 + \alpha/\pi$; a_i is the curvature radius in the point S_1 (S_2, S_3, S_4); θ_i is the angle between the wedge reference face and cone's axis. $\sigma(\theta_\perp) = \dfrac{8\pi a_0 L^2}{9\lambda}\left(1+\left(\dfrac{a_2-a_1}{L}\right)^2\right)^{3/2}$. where $a_0 = \left(\dfrac{a_2^{3/2}-a_1^{3/2}}{a_2-a_1}\right)^2$; $\theta_\perp = \dfrac{\pi}{2} - \operatorname{atan}[(a_2-a_1)/L]$.	In normal incidence to the generatrix Condition of normal incidence

Table 1.3 (continued)

No. 1	Reflector Type 2	Geometry Parameters 3	RCSs of Ideally Conducting Surfaces' "Bright" Elements 4	5
9.	Tip [4, Ch. 4, eq. 82]		$\sigma = \dfrac{\lambda^2 \tan^4 \alpha}{16\pi \cos^2 \theta}$ for $\theta \leq \alpha$	
10.	Disk		$\sigma(\theta) = \sigma_m \left\{ [\Lambda_1(2ka\sin\theta)]^2 + \left[\dfrac{J_2(2ka\sin\theta)}{ka} \right]^2 \right\}$, where $\sigma_m = 4\pi^3 a^4/\lambda^2$, $\Lambda_1(x) = 2\dfrac{J_1(x)}{x}$.	$a > 2\lambda$, $\|\theta\| < 80°$.

Table 1.3 (continued)

No. 1	Reflector Type 2	Geometry Parameters 3	RCSs of Ideally Conducting Surfaces' "Bright" Elements 4	5
11.	Luneberg lens	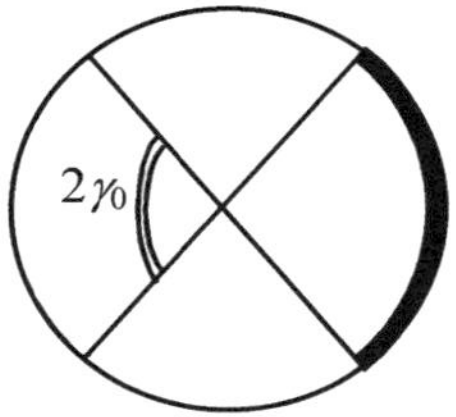	$\sigma_m = 4\pi^3 a^4/\lambda^4$	in sector $2\gamma_0$
12.	Surface traveling wave [4, Ch. 4, eqs. 72 and 73]		(see equations below)	(see notes below)

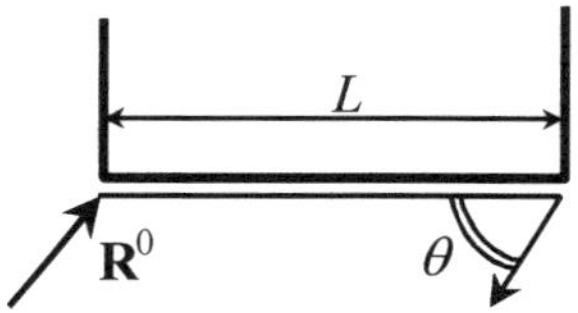

$$\sigma = \frac{\gamma^2\lambda^2}{\pi Q^2}\left\{\frac{\sin\theta}{1 - p\cos\theta}\sin\left[\frac{kL}{2p}(1 - p\cos\theta)\right]\right\}^4 \cos^4\Phi,$$

$$Q = -\frac{2}{p^2} + \frac{\operatorname{cin}[(kL/p)(1 + p)] - \operatorname{cin}[(kL/p)(1 - p)]}{p^3}$$

$$+ \frac{1}{2p^3}\left\{(p - 1)\cos[(kL/p)(1 + p)] + (p + 1)\cos[(kL/p)(1 - p)]\right.$$

$$\left. + (p^2 - 1)\frac{kL}{p}(\operatorname{si}[(kL/p)(1 + p)] - \operatorname{si}[(kL/p)(1 - p)])\right\}$$

Here, cin(x) is the modified integral cosine;
si(x) is the integral sine;
γ is the voltage reflection coefficient;
p is relative phase velocity;
L is the length of thin body (edge);
Φ is the angle between vector **E** and projection of body axis on the wave front

Column 5 notes (row 12):

p is defined as a ratio of the body length to the current path length along the surface;
If $p = 1$, $Q = \ln\left(\frac{4\pi L}{\lambda}\right) - 0.4228$;
$\gamma = 1/3$ for edges, thin rods, thin spheroid;
$\gamma = 0.7$ for the ogive;
Maximum of σ is observed if $\theta \approx 49.35\sqrt{\lambda/L}$ degrees.

Table 1.3 (continued)

No. 1	Reflector Type 2	Geometry Parameters 3	RCSs of Ideally Conducting Surfaces' "Bright" Elements 4	5
13.	Creeping wave for sphere [4, Ch. 4, eq. 71]		$\sigma \approx \pi a^2 1.03(ka)^{-5/2}$. Here, a is the sphere radius.	For ka varying from 1 to 15
14.	Specular interaction [4, Ch. 4, eq. 66]		$$\sigma = \frac{\pi \rho_{11}\rho_{12}\rho_{21}\rho_{22}}{d^2 \sin(2\gamma_1)\left[\sin(2\gamma_1) + \dfrac{\rho_{21}}{d}\cos\gamma_1 + \dfrac{\rho_{11}}{d}\sin\gamma_1\right]\left[2 + \dfrac{\rho_{12}}{d}\cos\gamma_1 + \dfrac{\rho_{22}}{d}\sin\gamma_1\right]},$$ where ρ_{ij} is the jth curvature radius of ith surface reflector.	Condition of interaction: $2\gamma_1 + 2\gamma_2 = \pi$, $\cos\gamma_1 = \lvert \mathbf{R}^0 \cdot \mathbf{n}_1 \rvert$, $\cos\gamma_2 = \lvert \mathbf{R}^0 \cdot \mathbf{n}_2 \rvert$.

Two types of bright elements—specular and edge elements—must be distinguished.

Specular elements, points, and lines are typical of smooth surfaces, such as quadric ones. Their positions are changed with a change of scattering direction (Table 1.3, items 1–3, 5, 7, 8, 11, 14, and partly 10). So, Table 1.1 contains expressions for the bright element coordinate calculation of eight ideally conductive smooth approximating surfaces. As it can be seen from Table 1.1, some objects have several bright points forming bright lines (items 4–8 of Table 1.3). Corresponding expressions for radii of curvature and other parameters are given in Table 1.2.

Edge bright points and lines have fixed positions on the body and are typical of objects with knife (Table 1.3, items 4–6 and 10 partly) and lance edges (Table 1.3, item 9).

In addition to Table 1.2, let us give some complementary considerations. The so-called stationary phase points define the bright points and lines. At each point $\mathbf{r}_\nu$ the front of a flat incident wave is a tangent to the convex surface $F_\nu(\mathbf{r}_\nu) = 0$ and normal to the vector $\mathbf{R}^0 = \left\| R_x^0, R_y^0, R_z^0 \right\|^{\mathrm{T}}$, where R_x^0, R_y^0, R_z^0 are direction cosines. The condition of colinearity of vectors grad F and $\mathbf{R}^0$ is

$$\frac{1}{R_x^0} \frac{dF}{dx_\nu} = \frac{1}{R_y^0} \frac{dF}{dy_\nu} = \frac{1}{R_z^0} \frac{dF}{dz_\nu} \tag{1.24}$$

Data from Table 1.2 satisfy (1.24) and the surface equation $F_\nu(\mathbf{r}_\nu) = F_\nu(x_\nu, y_\nu, z_\nu) = 0$.

1.3.6.1 Vector Transformations and Calculations

Transformation (1.12) through (1.18) of the incident wave's unit vector into the arbitrary νth local coordinate system precedes the use of initial data. Coordinate vectors $\mathbf{r}_i$ of each ith bright element are then calculated for all approximating surfaces $\nu = 0, 1, 2, \ldots, N_\Sigma - 1$ by means of data from Tables 1.1–1.3. Each of the isolated bright points and several points of bright lines are checked for membership in the appropriate νth target element and for absence of shadowing.

Checking for the ith bright point belonging to the νth target element is provided by verification of inequality $\Phi_{\nu k}(\mathbf{r}_i) < 0$, where $k = 0, 1, 2, \ldots, K_\nu - 1$.

1.3.6.2 Checking for Absence of Shadowing

It is provided by verification for absence of intersection of the line-of-sight $\mathbf{r}(s) = \mathbf{r}_i - s\mathbf{R}^0 = 0$, drawn from the radar to the ith bright point, with limited parts of other approximating surfaces of the second kind. If such intersection takes place, parameter s can be found as a real positive solution of the equation

$$\Psi_\gamma[\mathbf{r}(s)] = \mathbf{r}^T(s)\mathbf{P}_\gamma\mathbf{r}(s) = C \qquad (1.25)$$

where $\mathbf{P}_\gamma$ is the 3×3 matrix for coefficients of the canonical equation of the γth surface $\Psi_\gamma[\cdot] = C$, $C =$ const. All this leads to quadratic equation

$$as^2 + 2bs + c = 0, \text{ where } a = \mathbf{R}^{0T}\mathbf{P}_\gamma\mathbf{R}^0, \ b = \mathbf{R}^{0T}\mathbf{P}_\gamma\mathbf{r}_i, \ c = \mathbf{r}_i^T\mathbf{P}_\gamma\mathbf{r}_i - C$$

with solutions $s_{1,2} = (-b \pm \sqrt{b^2 - ac})/a$. Its real positive solution implies that the correspondent bright point is shadowed. Shadowing of bright lines and bright areas is verified individually, in steps Δl and ΔS, respectively.

1.3.6.3 PSM of Unshadowed Bright Points and Lines for Arbitrary Smooth Surfaces

The RCS of a bright point can be obtained for the short-wave case from the simple equation

$$\sigma = R_1 R_2 \qquad (1.26)$$

where $R_1 \gg \lambda$ and $R_2 \gg \lambda$ are the principal radii of curvature. Such radii for quadric surfaces can be found in Table 1.2. Since quadric surfaces have identical values of RCS σ_i in their own polarization basis, the PSM in this basis is $\mathbf{M}_i = \mathbf{diag}(\sqrt{\sigma_i}, \sqrt{\sigma_i}) = \sqrt{\sigma_i} \cdot \mathbf{I}$. Here, $\mathbf{diag}(a, b)$ is a diagonal matrix and $\mathbf{I}$ is the identity matrix. The PSM of ith bright point in the arbitrary polarization basis for $R_{1,2} \gg \lambda$ is

$$\mathbf{A}_i = \sqrt{\sigma_i}\,\mathbf{U}^{*T}\mathbf{I}\mathbf{U} = \mathbf{M}_i \qquad (1.27)$$

due to unitarity of matrix $\mathbf{U}$ in (1.6).

PSM of the unshadowed edge bright line is described by two unequal values of RCS in its own polarization basis; so, the PSMs can be found from the data in Tables 1.1 through 1.3 in the form of equation $\mathbf{M}_i = \mathbf{diag}(\sqrt{\sigma_{\|i}}, \sqrt{\sigma_{\perp i}})$.

Let us use a pair of unit vectors $\mathbf{l}^0$, $\mathbf{m}^0$, determined by (1.4), orthogonal to each other and to unit vector $\mathbf{R}^0$. Let us consider also the unit vector $\mathbf{l}^0_i$ parallel to the rib of the knife-edge surface. Transforming it with (1.15), (1.16)

$$\left\| l^0_{Xi} \quad l^0_{Yi} \quad l^0_{Zi} \right\|^{\mathrm{T}} = \mathbf{H}(\psi, \theta, \gamma)\mathbf{H}^{\mathrm{T}}(\psi_\nu, \theta_\nu, \gamma_\nu)\left\| l^0_{xi} \quad l^0_{yi} \quad l^0_{zi} \right\|^{\mathrm{T}} \tag{1.28}$$

into radar coordinate system, we can find the scalar product $((\mathbf{l}^0_i)^{\mathrm{T}}\mathbf{R}^0)$ and finally the product $((\mathbf{l}^0_i)^{\mathrm{T}}\mathbf{R}^0)\mathbf{R}^0$ that is the vector orthogonal to the wavefront (WF) plane. The difference $\mathbf{l}^0_i - ((\mathbf{l}^0_i)^{\mathrm{T}}\mathbf{R}^0)\mathbf{R}^0$ is the vector component of $\mathbf{l}^0_i$ vector in the wavefront plane (Figure 1.7). The angle φ_i between this component and vector $\mathbf{l}^0$ is determined by expression

$$\cos\varphi_i = \{\mathbf{l}^0_i - ((\mathbf{l}^0_i)^{\mathrm{T}}\mathbf{R}^0)\mathbf{R}^0\}\mathbf{l}^0 / \left| \mathbf{l}^0_i - ((\mathbf{l}^0_i)^{\mathrm{T}}\mathbf{R}^0)\mathbf{R}^0 \right| \tag{1.29}$$

Whether φ_i is positive or negative is determined by whether the scalar product of the determined vector component and vector $\mathbf{m}^0$ is positive or negative. Matrix $\mathbf{U}_i$ of polarization basis rotation and PSM $\mathbf{M}_i$ in the target's polarization basis are then used to evaluate PSM $\mathbf{A}_i$ in the radar polarization basis

$$\mathbf{A}_i = \mathbf{U}_i^{*\mathrm{T}}\mathbf{M}_i\mathbf{U}_i, \quad \mathbf{U}_i = \left\| \begin{matrix} \cos\varphi_i & -\sin\varphi_i \\ \sin\varphi_i & \cos\varphi_i \end{matrix} \right\| \tag{1.30}$$

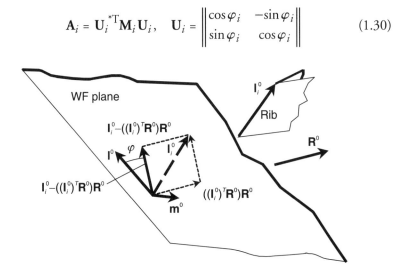

Figure 1.7 To the PSM of the unshaded edge specular line calculation.

1.3.6.4   Evaluation of the Whole Backscattered Signal

For scalar products $\mathbf{R}_i^{\mathrm{T}}(t)\mathbf{R}^0$ calculation, the coordinates of unshadowed bright points are transformed according to (1.13) into a common (target body's or radar's) coordinate system. For the target body's coordinate system, such calculation is provided by using (1.15), which allows the whole backscattered signal (1.20) to be evaluated.

1.3.7 Application Limits of the Simplest Component Simulation Method

The illustration (Figure 1.8) of several scattering mechanisms [2] is convenient in this case. The following mechanisms are shown: specular surface return, cavity return, interaction echo, edge diffraction, gap and seam echo, corner diffraction, tip diffraction, traveling and creeping wave echoes, and curvature discontinuity return. In the chosen frequency band—centimeter and short decimeter waves—the most essential mechanisms are as follows: specular surface return, interaction echo, edge diffraction, tip diffraction,

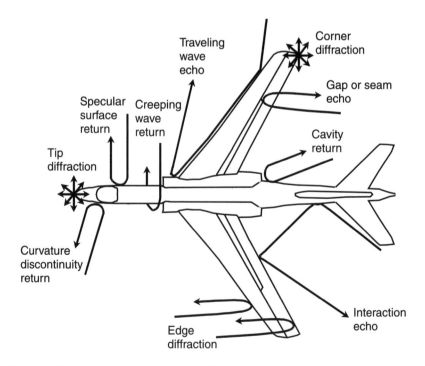

Figure 1.8 Several scattering mechanisms illustrated on the basis of the airframe of the Tu-16 aircraft (*After:* [2, Figure 1.6]).

corner diffraction that can be accounted for (as it was shown above), and cavity returns (including those of antennas and cockpits) that will be accounted for in Section 1.5 together with those of rotating structures. All these effects are considered in the program that will be issued after this book. The curvature discontinuity return was omitted due to its small contribution in the considered cases of nonstealth targets. Traveling and creeping waves (Table 1.3) are considered in the program only to the extent necessary to approximate simulation in the chosen frequency band. The model elaboration in the direction of using the longer waves remains a problem to solve. A comparison between using the simplest components method and some other methods of physical optics is given in Chapter 7.

1.4 Peculiarities of the Target Motion Simulation

The simulation of target motion has to take into account its influence on the backscattered signal (1.20). Target motion in the atmosphere depends on: (1) the kinematics of deterministic motion of the target mass center and target rotation relative to the line-of-sight (Section 1.4.1), and (2) the statistical characteristics of atmosphere and the dynamics of target-atmosphere interaction (Section 1.4.2). Only the kinematics of target mass center motion and the target orientation change connected with it can be used for the moderate bandwidth-duration product of signals or the calm atmosphere. Statistical dynamics of target yaws due to wind gusts limit coherence time of the signal, particularly that of very large duration and bandwidth. We will consider these questions in Section 1.4.2. Most of the simulation examples of target yaws' influence on signals will be given in Chapter 2 for the wideband signals and in Chapter 3 for narrowband ones. Statistical dynamics of the target mass center displacements (due to wind gusts) act on an arbitrary signal weaker than yaw and, therefore, will be neglected.

1.4.1 Deterministic Target Motion Description in Accounting for Earth's Curvature

Target mass centers' Cartesian coordinates X, Y, Z can be recalculated from those in radar spherical system using (1.12). The kinematics considered is described by the vector differential equation $d\mathbf{R}/dt = \mathbf{V}(t)$, where the velocity vector $\mathbf{V}(t)$ has only one nonzero component in the target velocity coordinate system (since vector $\mathbf{V}(t)$ has no diameter, its roll angle is meaningless). According to (1.12), (1.16), and (1.19) one can obtain

$$\frac{d}{dt}\left\| X(t) \quad Y(t) \quad Z(t) \right\|^{\mathrm{T}} = \mathbf{H}(0, \; -\varphi\cos\beta, \; \varphi\sin\beta)$$

$$\cdot \; \mathbf{H}(\psi - \delta', \; \theta - \delta'', \; 0)\left\| v(t) \quad 0 \quad 0 \right\|^{\mathrm{T}}, \quad (1.31)$$

$$\frac{d\varphi}{dt} = \frac{v(t)}{R_{\mathrm{ef}}}\cos(\psi - \delta')\cos(\theta - \delta'')$$

where δ' and δ'' are the aircraft's crab angle and angle of attack, which are normally small.

The target orientation is presented in the general case of unsteady movement by the following functions:

$$\psi = \psi(t), \; \theta = \theta(t), \; \gamma = \gamma(t) \tag{1.32}$$

or in the steady-state movement case by the following equations:

$$\psi = \text{const}, \quad \theta = \text{const}, \quad \gamma = \text{const}. \tag{1.33}$$

In the general case of unsteady movement, one can determine from (1.31) the Cartesian coordinate increments for small time intervals $t_n - t_{n-1}$ in the form

$$\left\| X_n - X_{n-1} \quad Y_n - Y_{n-1} \quad Z_n - Z_{n-1} \right\|^{\mathrm{T}}$$

$$\approx \mathbf{H}(0, \; -\varphi_{n-1}\cos\beta_{n-1}, \; \varphi_{n-1}\sin\beta_{n-1}) \tag{1.34}$$

$$\cdot \; \mathbf{H}(\psi_{n-1} - \delta'_{n-1}, \; \theta_{n-1} - \delta''_{n-1}, \; 0)\left\| \nu_{n-1} \quad 0 \quad 0 \right\|^{\mathrm{T}}(t_n - t_{n-1})$$

The small Cartesian coordinates' increments $X_n - X_{n-1}$, $Y_n - Y_{n-1}$, $Z_n - Z_{n-1}$ can be recalculated into spherical ones $R_n - R_{n-1}$, $\beta_n - \beta_{n-1}$, $\epsilon_n - \epsilon_{n-1}$ after introducing the recalculation matrix $\mathbf{G}(R, \; \beta, \; \epsilon)$. Using (1.12), we obtain

$$\mathbf{G}(R, \; \beta, \; \epsilon) = \left\| \begin{matrix} \partial X/\partial R & \partial X/\partial\beta & \partial X/\partial\epsilon \\ \partial Y/\partial R & \partial Y/\partial\beta & \partial Y/\partial\epsilon \\ \partial Z/\partial R & \partial Z/\partial\beta & \partial Z/\partial\epsilon \end{matrix} \right\| \tag{1.35}$$

$$= \left\| \begin{matrix} \cos\epsilon\cos\beta & -R\cos\epsilon\sin\beta & -R\sin\epsilon\cos\beta \\ \sin\epsilon & 0 & R\cos\epsilon \\ \cos\epsilon\sin\beta & R\cos\epsilon\cos\beta & -R\sin\epsilon\sin\beta \end{matrix} \right\|$$

Applying the inversion of the recalculation matrix $\mathbf{G}(R,\ \beta,\ \epsilon)$, we obtain the approximate recurrent equation, which determines the current target range R_n and radar-target line of sight β_n, ϵ_n orientation,

$$\left\| R_n - R_{n-1}\quad \beta_n - \beta_{n-1}\quad \epsilon_n - \epsilon_{n-1} \right\|^{\mathrm{T}}$$
$$= \mathbf{G}^{-1}(R_{n-1},\ \beta_{n-1},\ \epsilon_{n-1})\mathbf{H}(0,\ -\varphi_{n-1}\cos\beta_{n-1},\ \varphi_{n-1}\sin\beta_{n-1})$$
$$\cdot\, \mathbf{H}(\psi_{n-1} - \delta',\ \theta_{n-1} - \delta'',\ 0)\left\| v_{n-1}\quad 0\quad 0 \right\|^{\mathrm{T}}(t_n - t_{n-1})$$

$$(1.36)$$

$$\varphi_n - \varphi_{n-1} = \frac{V(t_n - t_{n-1})}{R_{\mathrm{ef}}}\cos(\psi_{n-1} - \delta'_{n-1})\cos(\theta_{n-1} - \delta'')$$

$$(1.37)$$

Equation (1.36) can be used both for unsteady- and steady-state target motion.

1.4.2 Statistical Properties of Atmosphere and Dynamics of Target-Atmosphere Interaction

The moving target observes a wind opposite to its velocity vector $\mathbf{V}$ even in the calm atmosphere. Turbulent atmosphere heterogeneity described by its statistics leads to the wind gust formation and angular target yaws as a result of the target-atmosphere interaction. Yaw of the target moving with the mean velocity described by vector $\mathbf{V}$ is caused by the wind gust component $\Delta\mathbf{V}_\perp$, orthogonal to $\mathbf{V}$. As it is shown in Figure 1.9, an angle μ occurs between the air flux direction and the vector $(-\mathbf{V})$ direction. A torque rotating the target arises in order to decrease the angle μ. In addition to the target steady-state crab angle δ' and angle of attack δ'', yaw occurs because of the changing wind gust components $\Delta\mathbf{V}_\perp$ along the flight path. Having been worked out by the target control system that includes the pilot or autopilot,

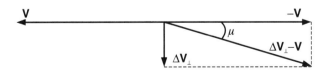

Figure 1.9 To the formation of an angle between the direction of the blowing air flux and vector $(-\mathbf{V})$, opposite to the target velocity, due to the action of wind gust component $\Delta\mathbf{V}_\perp$, transverse to vector $(-\mathbf{V})$.

the yaw is decreased. But this decrease is incomplete, and remaining yaw acts on the backscattered extended signals.

The block diagram of yaw simulation (Figure 1.10) is used separately for each of the Cartesian coordinates. Its first element (Figure 1.10) is the sample generator of white Gaussian noise of unit spectral density. The second element is a dynamic unit of the second order with the transfer function $K_{atm}(p)$, which simulates wind gust components $\Delta V_\perp$ taking into account their spatial correlation in the atmosphere. The third element is a dynamic unit of zero order with a transfer function $K_\mu(p) = 1/V$ = const describing the angle μ between the directions of air flux and vector ($-V$). The fourth element is a dynamic unit of second order with a transfer function $K_{tg}(p)$, taking into account peculiarities of aerodynamic forces acting on the target, the target's inertial characteristics, and the smoothing effect of the pilot or autopilot.

Wind gust simulation is provided on the basis of the transfer function [17]:

$$K_{atm}(p) = \sqrt{\frac{3}{a\Delta t}} \cdot \sigma_{\Delta v} \cdot \frac{p + \dfrac{1}{a\sqrt{3}}}{p^2 + \dfrac{2p}{a} + \dfrac{1}{a^2}} \tag{1.38}$$

where $a = L/V$ is the ratio of the atmosphere turbulence scale L to the mean target velocity V, and $\sigma_{\Delta v}$ is the wind gust velocity standard deviation. Estimated mean and maximum values of turbulence scale are $L_{mean} \approx 400m$ and $L_{max} \approx 600m$ for altitudes of 3 to 7 km and $L_{mean} \approx 990m$ and $L_{max} \approx 2600m$ for altitudes of 15–18 km [18]. Estimated values of velocity standard deviation $\sigma_{\Delta v}$ are shown in Table 1.4 [18, 19].

As for the turbulence scale, these parameters are approximate and dependent on altitude. For instance, for clear weather and altitudes of 3 to 7 km, the maximum standard deviation of wind gusts is 2 m/s and its mean value is 0.46 m/s; for altitudes of 15 to 18 km the maximum standard deviation of wind gusts is 1.69 m/s and its mean value is 0.86 m/s [18].

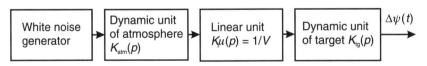

Figure 1.10　Block diagram of the target course yaw $\Delta\psi$ simulation. Similar block diagrams are used to simulate the pitch $\Delta\theta$ and roll $\Delta\gamma$ yaws.

Table 1.4
Estimated Values of Standard Deviations of Wind Gusts

Weather Conditions of Atmosphere Turbulence	Standard Deviation of Wind Gust's Velocities
Practically quiet atmosphere turbulence	$\sigma_{\Delta V} < 0.5$ m/s
Clear weather turbulence	$0.5 < \sigma_{\Delta V} < 2$ m/s
Cloudy weather turbulence	$2 < \sigma_{\Delta V} < 4$ m/s
Stormy weather turbulence	$\sigma_{\Delta V} > 4$ m/s

Simulation of the angle $\mu = \mu(t)$ is provided by an inertialess dynamic unit of zero order with transfer function $K_\mu(p) = 1/V$. The division operation $\mu(t) = |\Delta V_\perp(t)|/V$ provides recalculation of wind gusts normal to velocity of target movement into angle deflection of the target orientation (Figure 1.9).

Simulation of target reaction of the angle $\mu = \mu(t)$ is provided on the basis of transfer function [17]

$$K_{tg}(p) = -\omega_{c.off} \frac{p - 2/\tau_{p.a}}{p^2 - (\omega_{c.off} - 2/\tau_{p.a})p + 2\omega_{c.off}/\tau_{p.a}} \quad (1.39)$$

Here, $\omega_{c.off}$ is the cut-off frequency of logarithmic amplitude-frequency response, and $\tau_{p.a}$ is a time constant of the pilot-target or autopilot-target system. Estimated values of $\omega_{c.off}$ and $\tau_{p.a}$ are shown in Table 1.5.

Simulated yaws of ψ and θ target motion parameters can be included in (1.36) to consider their influence on the backscattered signal. Some experimental data about the intervals of backscattered signal coherence are given in [20].

Table 1.5
Cut-Off Frequencies and Time Constants for the Pilot-Target and Autopilot-Target Systems

Target Class	Time Constant of the Target Pilot or Autopilot, $\tau_{p.a}$ (sec)	Cut-Off Frequency, $\varphi_{c.off}$ (rad /sec)
Large-sized	0.3	1.5–2
Medium-sized	0.3	2.5–3
Small-sized	0.1	3.5–4

1.5 Peculiarities of Simulation of Fast Rotating Elements

Rotating elements of targets act on backscattered signals causing modulation. Examples of rotational modulation simulation will be given in Chapter 2 for wideband signals and in Chapter 3 for narrowband ones. Here, we will consider the essence of rotational modulation of scattered signals (Section 1.5.1), the simulation specific to jet-engine modulation (JEM, Sections 1.5.2 and 1.5.3) and propeller modulation (PRM, Section 1.5.4) [21–29].

1.5.1 Essence of Rotational Modulation of Scattered Signals

Origins of rotational modulation for various aerial targets are their rotating systems, such as the blades of propellers and jet engines' compressors and turbines. The signal scattered by a rotating system acquires distinctive rotational modulation, which influences the target detection and hinders velocity tracking of extended coherent signals, but which is useful as a target recognition feature.

Rotational modulation produced by aerial targets depends on their type and construction, as well as on the wavelength. Missiles do not usually produce rotational modulation at decimeter and centimeter waves. Slow rotor modulation is specific to helicopters.

Somewhat faster rotational modulation is specific to the propellers of turbo-prop aircraft with gas-turbine (turbo-prop) engines used at subsonic speeds. A gas-turbine engine usually has an air compressor either on the propeller rotation axis or on a separate one. The compressor's rotational modulation is perceptibly faster than that of the propeller.

Very fast rotational modulation is specific to aircraft with turbo-jet engines (jet aircraft). The compressor (air-scoop, fan) is located at the front of the engine, while the turbine is at its rear.

More common than single-stage turbines or compressors with a whole number of blades N are multistage ones. Both first and second stages with N_1 blades in the first stage and $N_2 = N - N_1$ blades in the second one can cause noticeable rotational modulation. Multi-engine aircraft that are safer in operation than single-engine ones are chiefly used. Therefore, consideration of multi-engine and multistage rotating systems is essential. The form of the radar illumination (continuous, time-restricted, or bursts of pulses with various repetition frequencies) and the bandwidth also influence the observed effect of rotational modulation.

Rotational Modulation of Sinusoidal Illumination Signals Caused by Single-Stage and Single-Engine Rotating Systems. Assuming identical blades, rotational modulation corresponds to the periodical change of sinusoidal signal amplitude and phase with a period $1/NF_{rot}$. Here, N is the number of blades and F_{rot} is the rotation frequency. For independent scattering by each of the blades, the instantaneous value of scattered signal is defined by the following equation:

$$E(t) \equiv \sum_{\nu=0}^{N-1} \sqrt{\sigma_b\left(t - \frac{\nu}{NF_{rot}}\right)} \qquad (1.40)$$

$$\cdot \cos\left\{2\pi f_0 t + C_\nu \sin\left[2\pi F_{rot}\left(t - \frac{\nu}{NF_{rot}}\right) + \phi_\nu\right]\right\}$$

Here, $\sigma_b(s)$ is the blade RCS as a function of time, f_0 is the carrier frequency, ϕ_ν are the constant initial phases (depending on the choice of the time reference origin), and C_ν are constants depending on ν. The frequency spectrum of any periodical function is a line spectrum with the spectral lines at the frequencies kNF_{rot}, $k = 0, \pm 1, \pm 2, \ldots, k_{max}$ [Figure 1.11(a)] and the maximum interval NF_{rot} between the lines. Some harmonics of an amplitude-phase modulated signal spectrum can vanish. The possible k_{max} value for JEM is considered in Section 1.5.3.

Rotational Modulation of Time-Restricted Sinusoidal Illumination Signals Caused by Single-Stage and Single-Engine Rotating Systems. A sinusoid with rotational modulation (1.40) is multiplied by a video signal $U(t)$ of limited duration T_0, so that the convolution of their spectrums takes place. We can consider this case as the modulation of each harmonic of (1.40) by $U(t)$. The spectral lines of rotational modulation become blurred into the spectral regions of extent $1/T_0$ [Figure 1.11(b)] for the condition of $T_0 F_{rot} \gg 1$.

Rotational Modulation of a Burst of Sinusoidal Pulses Caused by Single-Stage and Single-Engine Rotating Systems. Sinusoid with rotational modulation (1.40) is multiplied by a burst video signal with a pulse repetition frequency (PRF) equal to F_{pr}. For infinite signal duration $T_0 \to \infty$, the spectral lines appear on combinational frequencies $k_{pr} F_{pr} + k F_{rot}$, where $k_{pr} = 0, \pm 1, \pm 2, \ldots, \pm \mu$. For finite signal duration T_0, these lines become blurred into spectral regions of extent $1/T_0$ [Figure 1.12]. The spectrum shown for the high PRF case [Figure 1.12(a)] is identical to that for CW illumination [Figure 1.11(b)], but with aliasing at the PRF. For medium or low PRF the

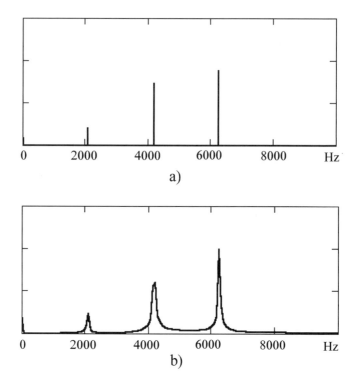

Figure 1.11 Simulated rotational modulation spectrum caused by single-stage and single-engine rotating system for (a) continuous sinusoidal illumination signal on the wavelength of 3 cm; and (b) restricted in time one for the signal duration of about 12 ms. External and internal blade radii are 0.54m and 0.13m, respectively; number of blades is 27; rotation frequency is about 120 Hz.

spectrum of Figure 1.11(b) is reproduced with medium or high distortion caused by aliasing.

Rotational Modulation Caused by Multiengine Rotating System. If rotation frequencies F_{rot} of all engines are identical, superposition of the engines' backscattering does not distort the rotational modulation. Otherwise, there is beating of the extended signals backscattered from engines and an increase in the number of spectral lines (their duplication in the [29, Figure 7] experiment).

Rotational Modulation Caused by Multistage Rotating System. It creates, in the CW case, spectral lines at the combinational frequencies, for example,

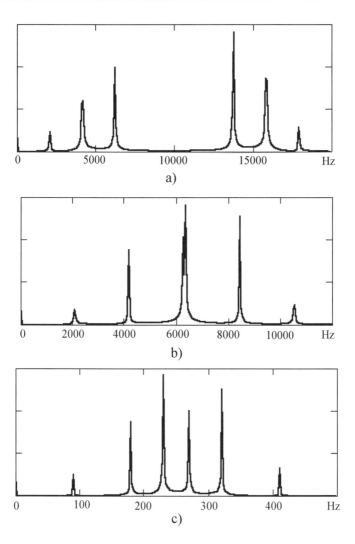

Figure 1.12 Simulated rotational modulation spectrum of the pulse burst for: (a) high PRF; (b) medium PRF; (c) low PRF caused by single-engine and single-stage rotational system.

at the frequencies $(k_1 N_1 + k_2 N_2)F_{rot}$, $k_{1,2} = 0, \pm1, \pm2, \ldots$ in a two-stage case. Lines associated with the second stage $k_2 \neq 0$ are weaker than those of first-stage $k_2 = 0$. In periodic pulsed (burst) illumination the spectrum becomes additionally complicated. For the two-stage case, combinational spectral lines are created at the frequencies $k_{pr}F_{pr} + (k_1 N_1 + k_2 N_2)F_{rot}$, blurred into the spectral regions of extent $1/T_0$ (Figure 1.13). The spectral lines of Figure 1.13 are similar to the experimental ones of [29, Figure 6] for the high PRF.

PRF Limitations of the Rotational Information. They do not lead to modulation distortions if the PRF $F_{pr} > 2 (F_{rmax} + 1/T_0)$. Here, F_{rmax} is the maximum absolute value of (1) $(k_1 N_1 + k_2 N_2) F_{rot}$ for the two-stage rotating system and (2) kNF_{rot}, for the one-stage system (or two-stage one neglecting weak spectral elements). If target spectral components are concentrated in a bandwidth ΔF_r, the condition $F_{pr} > 2(\Delta F_r + 1/T_0)$ for absence of distortion can be used. Such evaluations are rough; more detailed analysis is given in Chapter 3.

Wave Band Limitations on Rotational Information. Nonuniform waveguides are formed between the blades, and rotational modulation occurs if radio waves penetrate into them. The condition for wave penetration can be roughly expressed by the known inequalities

$$2\pi R_{ext}/N > \lambda/2 \quad \text{or} \quad f > f_{crit} = cN/4\pi R_{ext} \qquad (1.41)$$

Here, R_{ext} is the exterior radius of the rotating structure, N is the number of blades, and f_{crit} is the critical frequency of the nonuniform waveguide's entrance. Engine rotational modulation emerges therefore at centimeter and short decimeter waves only, whereas propeller modulation can be observed in the broad waveband from centimeter to meter, and even decameter, waves.

Tentative parameters of turbojet engines of aerial targets used in simulation are shown in Table 1.6.

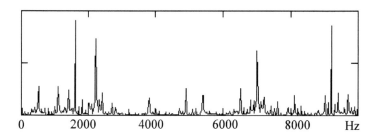

Figure 1.13 Simulated rotational modulation spectrum for a burst signal on the wavelength of 3 cm caused by two-stage rotational system. External and internal blade radii are 0.54m and 0.13m, respectively; number of blades in first and second stages are 27 and 43, respectively; rotation frequency is about 120 Hz.

Table 1.6
Tentative Parameters of Compressors and Turbines of the Simulated Target Engines

Class of Target		External Diameter (m)	Internal Diameter (m)	Parameter of Engine's Compressor and Turbine			Rotation Frequency (Hz)	Blade's Angle of Attack (degree)
				Large Blade Size (m)	Small Blade Size (m)	Number of Blades		
Large-sized turbo-jet	Comp.	0.9–1.2	0.2–0.7	0.2–0.35	0.1–0.15	28–36	70–120	18–20
	Turb.	0.9–1.3	0.3–0.8	0.2–0.25	0.06–0.08	68–75	70–120	20–30
Medium-sized turbo-jet	Comp.	0.7–0.9	0.25–0.5	0.2–0.3	0.06–0.08	30–35	160–180	15–20
	Turb.	0.7–0.9	0.25–0.6	0.1–0.15	0.04–0.05	50–75	160–180	20–30
Missile	Comp.	0.2–0.3	0.08–0.1	0.07–0.09	0.01–0.03	30–40	500–520	10–20
	Turb.	0.2–0.3	0.1–0.15	0.05–0.07	0.01–0.03	70–80	500–520	20–30

1.5.2 Simulation of JEM Neglecting Shadowing Effects

Jet-engine modulation is caused by rotation of the fans, compressors, and turbines. The peculiarity of all these rotating elements is the large number of identical blades with definite angles of attack. Their mutual shadowing and presence of the air-scoop are neglected here and will be considered in Section 1.5.3. Rotational modulation of continuous sinusoidal signals only is examined in Sections 1.5.2 and 1.5.3, while pulsed signal modulation has been considered in Section 1.5.1.

1.5.2.1   The Asymptotic Case of Backscattering from a Stationary Single Blade

This makes possible the use of physical optics without repeated reflections, typical of waveguides. The validity of inequality $f >> f_{\text{crit}}$ is supposed, which is stricter than inequality (1.41). The νth blade, $\nu = 0, 1, 2, \ldots, N - 1$ is described approximately by a rectangular ideal conductive plate $x_\nu = 0$, $R - b/2 \leq z_\nu \leq R + b/2$, $-a/2 \leq y_\nu \leq a/2$ in its own local coordinate system $O_\nu x_\nu y_\nu z_\nu$, where R is the radius of blade centers' position (Figure 1.14).

     The νth blade local coordinate system $O_\nu x_\nu y_\nu z_\nu$ has been rotated with respect to the target system $O_{\text{tg}} \xi \eta \zeta$:

1. Around the axis $O_{\text{tg}} \zeta$ by the blade angle of attack θ_0 independent of ν;

2. Around the axis $O_{\text{tg}} \xi$ by the blade roll angle $\gamma_\nu = 2\pi\nu/N$ depending on ν.

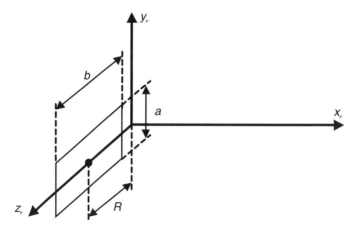

Figure 1.14 Turbine blade in local coordinate system approximated by rectangular plate.

The origin of the local coordinate system $O_\nu x_\nu y_\nu z_\nu$ is displaced from that of the target $O_{\text{tg}}\xi\eta\zeta$ by the radius vector $\boldsymbol{\rho}_\nu = \boldsymbol{\rho}_0 = \left\| \xi_0 \quad \eta_0 \quad \zeta_0 \right\|^{\text{T}}$, where ξ_0, η_0, ζ_0 are coordinates of the rotating system center.

Let us evaluate the signal backscattered by the νth blade. The propagation unit vector $\mathbf{R}^0$ of the incident wave must be recalculated from the radar coordinate system into the target one and then into the local one; then we have

$$\left\| R^0_{x\nu} \quad R^0_{y\nu} \quad R^0_{z\nu} \right\|^{\text{T}} \tag{1.42}$$

$$= \mathbf{H}^{\text{T}}(0, \ \theta_0, \ \gamma_\nu)\mathbf{H}^{\text{T}}(\psi, \ \theta, \ \gamma)\left\| \cos\epsilon\cos\beta \quad \sin\epsilon \quad \cos\epsilon\sin\beta \right\|^{\text{T}}$$

According to physical optics and (1.3) and (1.6), the diagonal elements of the blade polarization matrices are

$$\sqrt{\sigma_\nu}e^{j\varphi_\nu} = \frac{2\sqrt{\pi}}{\lambda} \int\limits_{-a/2}^{a/2} \int\limits_{R-b/2}^{R+b/2} \exp\left[-\frac{j4\pi}{\lambda}(R^0_{y\nu}y_\nu + R^0_{z\nu}z_\nu) \right] R^0_{x\nu}dy_\nu dz_\nu \tag{1.43}$$

Equation (1.43) adds together the reflections from the plate elements, each with the area $dy_\nu \cdot dz_\nu$ and the radius vector $\mathbf{r}_\nu = y_\nu\mathbf{y}^0_\nu + z_\nu\mathbf{z}^0_\nu$, where $\mathbf{y}^0_\nu$ and $\mathbf{z}^0_\nu$ are unit vectors of the local coordinate system. The sum $R^0_{y\nu}y_\nu + R^0_{z\nu}z_\nu$ that entered in the index of a power of (1.43) characterizes the pathlength difference for this element, formed by the scalar product $\mathbf{r}^{\text{T}}_\nu\mathbf{R}^0 = R^0_{y\nu}y_\nu + R^0_{z\nu}z_\nu$. The product $R^0_{x\nu}\,dy_\nu \cdot dz_\nu$ characterizes the projection of elementary area $dy_\nu \cdot dz_\nu$ onto the plane normal to unit vector $\mathbf{R}^0$ of the incident wave that is described by the scalar product $(dy_\nu \cdot dz_\nu\mathbf{x}^0_\nu)^{\text{T}}\mathbf{R}^0 = R^0_{x\nu}\,dy_\nu \cdot dz_\nu$, where $\mathbf{x}^0_\nu$ is the third unit vector of the local coordinate system. After integration in (1.43), one can obtain the polarization matrix diagonal element in the physical optics approximation

$$\sqrt{\sigma_\nu}e^{j\varphi_\nu} = \frac{2\sqrt{\pi}}{\lambda}ab\,\frac{\sin(2\pi aR^0_{y\nu}/\lambda)}{2\pi aR^0_{y\nu}/\lambda}\,\frac{\sin(2\pi bR^0_{z\nu}/\lambda)}{2\pi bR^0_{z\nu}/\lambda}\,R^0_{x\nu}e^{-j4\pi RR^0_{z\nu}/\lambda} \tag{1.44}$$

1.5.2.2   The Asymptotic Case of Backscattering from Single-Stage and Single-Engine Rotating Systems

This case corresponds to independent reflections from each blade neglecting their mutual shadowing. To account for the νth blade rotation, one can modify the expression of the blade roll angle in the target coordinate system $O_{\mathrm{tg}}\xi\eta\zeta$

$$\gamma_\nu = \gamma_\nu(t) = 2\pi F_{\mathrm{rot}}t - 2\pi\nu/N = 2\pi F_{\mathrm{rot}}(t - \nu/NF_{\mathrm{rot}}) \quad (1.45)$$

The instantaneous value $E(t)$ of the continuous sinusoidal signal reflected by the identically rotating blades of the single-stage compressor or turbine for $f \gg f_{\mathrm{cr}}$ is

$$E(t) \equiv \mathrm{Re}\left[\sum_{\nu=0}^{N-1}(\sqrt{\sigma_\nu}e^{j\varphi_\nu})e^{-j\varphi}\right] \quad (1.46)$$

Here,

$$\varphi = 2\pi(\mathbf{R}_{\mathrm{tg}} + \boldsymbol{\rho}_{\mathrm{en}})^{\mathrm{T}}\mathbf{R}^0/\lambda \quad (1.47)$$

where $\mathbf{R}_{\mathrm{tg}}$ is the radius vector of the target coordinate system origin in the radar one, and $\boldsymbol{\rho}_{\mathrm{en}}$ is the radius vector of the engine center position (local coordinate system origin) in the target coordinate system.

According to (1.42), the component $R^0_{z\nu}$ of propagation unit vector is a function of angle $\gamma_\nu = \gamma_\nu(t)$ and determines rotational phase modulation. For every combination of angles ψ, θ, γ and given parameters R, a, b, θ_0, (1.42) through (1.46) allow estimating the rotational modulation character. The presence of blade's angle of attack $\theta_0 \neq 0$ leads to spectrum asymmetry.

Departure from the physical optics and depolarization effects in JEM simulation arise when one of the blade's dimensions becomes comparable or small relative to the wavelength. The induced current and backscattering intensity are changed, especially if the electric field vector is oriented along the short blade's side $a < b$. Heuristically, this changing can be evaluated by means of the "long transmission line with losses" model that is open at both sides and fed from its middle. The whole relative conductivity of such a model is proportional to

$$Y_{\mathrm{a}} = j\tan\left[\left(\frac{2\pi}{\lambda} - j\alpha\right)\frac{a}{2}\right]$$

where the parameter α considers the losses in the line. Evaluating the tangent of a complex argument, we have [30, eq. 4.3.57]

$$Y_a = j\frac{\sin(2\pi a/\lambda) - j\sinh(\alpha a)}{\cos(2\pi a/\lambda) + \cosh(\alpha a)} \qquad (1.48)$$

For ($\alpha a \gg 1$ (i.e., $\sinh(\alpha a) \gg 1$, $\cosh(\alpha a) \approx \sinh(\alpha a) \gg 1$) we have $Y_a \approx 1$. It means that we may not take into account the edge effects for great blade size a. Conversely, for values $a = \lambda/4$ the resonance effect takes place. If $\alpha a \ll 1$, we have $Y_a = j\tan(\pi a/\lambda)$, and then $a \ll \lambda$; the value $Y_a \approx \pi a/\lambda$, and this case corresponds to Rayleigh scattering. Results of (1.48) for variation of the physical optics coefficient $|Y_a|$ with the ratio a/λ are given in Figure 1.15 for two values of the product $\alpha\lambda = 2$ (solid line) and $\alpha\lambda = 4$ (dashed line).

The discussion presented above shows that the following equation can be used as an improvement on the physical optics approximation of the blade PSM in its own polarization basis:

$$\mathbf{M}_\nu = \left\|\begin{matrix} \sqrt{\sigma_\nu}|Y_b| & 0 \\ 0 & \sqrt{\sigma_\nu}|Y_a| \end{matrix}\right\| \qquad (1.49)$$

Here, $|Y_b|$ is determined by (1.48) after substitution for the blade size a with size b and taking the absolute value. Equation (1.49) may be used also in the nonasymptotic case, when $|Y_a| = |Y_b| = 1$. It is supposed also that product $\alpha\lambda$ is evaluated heuristically or on the experiment base.

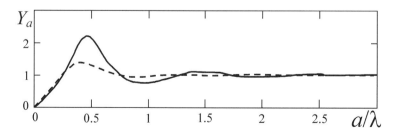

Figure 1.15 Coefficient $|Y_a|$ via a/λ ratio for the products $\alpha\lambda = 2$ (solid line) and $\alpha\lambda = 4$ (dashed line).

1.5.3   JEM Simulation, Taking into Account the Shadowing Effect and Related Topics

JEM Asymptotic Simulation, Taking the Shadowing Effect into Account. This can be performed by different methods. One or several bright points, bright lines, or bright areas (see Section 1.3.6) can replace each νth scattering element.

Let us begin with segregation of bright points from the solution (1.44) of Section 1.5.2. Each sine function of (1.44) or one of them can be transformed by the Eulerian formula $\sin\varphi = (e^{j\varphi} - e^{-j\varphi})/2j$. Each of N illuminated blades can introduce four or two summands in (1.44) that correspond to four angle bright points or two specular lines approximating the blade. The summands, corresponding to the shaded bright points or lines, must be considered carefully. It is better to divide the plane surfaces into parts and check these parts for the shadowing instead of edge specular points. Summation of fields scattered by such unshadowed areas replaces integration in (1.43). Approximate narrowing of the plate width caused by shadowing is another simple variant for solution.

JEM Asymptotic Simulation, Taking the Air-Scoop Presence into Account. The air-scoop is made as a rectangular or cylindrical duct through which the compressor or fan are fed with air. An exact solution of electromagnetic wave guiding in this duct is complicated. Therefore, let us be restrictive with a qualitative analysis of this subject. Figure 1.16 shows a case of electromagnetic wave propagation to the rotating system between two conducting planes placed apart from each other, the distance being many times larger than the wavelength. The main backscattering from the rotating system is observed at the same angle as in free space. A part of the backscattered energy can be

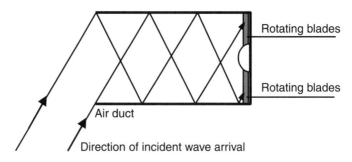

Figure 1.16   The case of electromagnetic wave propagation to a rotating system between two conducting planes placed apart from each other, the distance being many times larger than the wavelength.

lost; additional backscattering is possible due to the absence of electrodynamic matching. The rough evaluation of the RCS is then equal to $A\sigma_{rot} + B$. Values of coefficients A and B depend on angles of incidence, and they are selected empirically.

JEM Nonasymptotical Simulation. Functions described by (1.40) and (1.44) through (1.46) are periodic with a period $1/NF_{rot}$ corresponding to the displacement of each blade or interblade distance. As in (1.40), a Fourier series can be developed with frequencies kNF_{rot}, where $k = 0, \pm1, \ldots, \pm\mu$. The practical number of these harmonics depends on:

1. Degree of angularity or evenness of electromagnetic field at the entrances of interblade waveguides;
2. The blade angle of attack.

For JEM the number of half-waves between the blades tentatively defines the number $\mu = (2\pi R/N){:}(\lambda/2) = 4\pi R/N\lambda$.

Simulation of Multiengine and Multistage Rotating Systems. The number of the summands in calculation of equations similar to (1.46) increases in both kinds of systems. Various locations $\boldsymbol{\rho}_{en} = \boldsymbol{\rho}_l$ of rotating structures $l = 0, 1, 2, \ldots, L$ in (1.46) through (1.47) can be accounted for.

Simulation of only two-stage ($L = 2$) rotating structures with combinational frequencies $(k_1 N_1 + k_2 N_2)F_{rot}$ can be considered as adequate because of wave attenuation in rotating stages (see Figure 1.13 and experiment [29]). But such simulation will be effective only in cases of detailed engine description and carrying through of complicated calculations. It is simpler to realize an approximate multiplication of the one-stage simulation result by the heuristic value $\alpha + \beta\cos(2\pi k_2 N_2 t + \phi)$, where the sum $\alpha + \beta$ is of somewhat greater unity.

1.5.4 Simulation of PRM

The propeller as a rotating system differs from the turbine in that there is a smaller number of identical blades N and a smaller rotating frequency F_{rot}. According to (1.40) and (1.41), the PRM frequency spectrum differs from that of JEM by the smaller interval between spectral lines (regions) and by their increased number. More free access of radio waves to the blades is available. Therefore, the effect of shadowing of one blade by others can be completely neglected in simulation. Several approaches exist to approximate the propeller blade: with a flat plate, with a twisted plate, and with some of the simplest components. Let us consider and compare these approximations.

Propeller blade's approximation with a flat plate is carried out similarly to the analogous approximation (1.44) through (1.47) of a turbine's blade. Both phase modulation (due to the blade's phase center rotational motion) and amplitude modulation (due to the change of blade's orientation relative to the radar) are considered in (1.44) through (1.47). Depolarization effects can also be considered using equations (1.48) and (1.49).

Propeller blade's approximation with a twisted plate accounts for the blade twist along its longest axis. Such a twist $\theta_0 = \theta_0(z) = A - Bz$ in (1.42) leads to shifts of specular point positions on a blade in the process of its rotation and to additional phase modulation, widening the modulation spectrum [27]. Replacing the ith twisted blade by a set of small plates with unequal angles of attack, one can obtain the integral of (1.44) type in limits of variable z from L_1 to L_2:

$$\sqrt{\sigma_\nu}e^{j\varphi_\nu} = \frac{2\sqrt{\pi}}{\lambda}a\int_{L_1}^{L_2}\frac{\sin(2\pi aR_{y\nu}^0(z)/\lambda)}{2\pi aR_{y\nu}^0(z)/\lambda}R_{x\nu}^0(z)e^{-j4\pi zR_{z\nu}^0}dz \quad (1.50)$$

Assuming $(\sin x)/x \approx 1$ due to the relatively small a size, one can obtain

$$\sqrt{\sigma_\nu}e^{j\varphi_\nu} = \frac{2\sqrt{\pi}}{\lambda}a\int_{L_1}^{L_2}R_{x\nu}^0(z)e^{-j4\pi zR_{z\nu}^0}dz \quad (1.51)$$

For propeller blade's approximation with the simplest components, each blade can be considered on the basis of being a very thin sharp-edged ellipsoid (Figure 1.17). The blade's angle of attack is the angle between the rotational plane and the medium semiaxis of ellipsoid. The largest semiaxis of ellipsoid is equal to half the blade length, its medium semiaxis is equal to half the blade breadth, and its smallest semiaxis is equal to half the blade thickness [28]. The largest and medium semiaxes are much greater than the wavelength, and the smallest one is not. O. I. Sukharevsky proposed to complement the ellipsoid in its largest section by a wedge with a curved rib, tangent to the ellipsoid surface (see Section 7.1.8). Blade backscattering can be computed then as the physical optics reflection from the visible part of

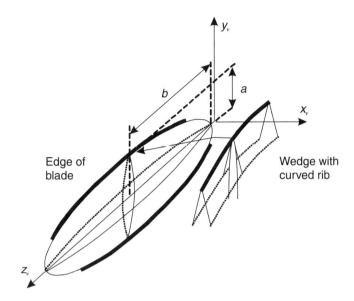

Figure 1.17 Propeller blade approximation with ellipsoid and curved rib wedge.

ellipsoid, from the edge specular point or their superposition. Only one specular point of the ellipsoid is visible. Its RCS can be expressed as $\sigma_e = \pi \rho_1 \rho_2$, where ρ_1 and ρ_2 are the principal radii of curvature at this point. Coordinates of the ellipsoid's specular points are given in Table 1.1 (item 1); its principal radii of curvature can be obtained from Table 1.2. (item 1). Two specular points of the edge, symmetrically situated, can be visible in general. The front point is always illuminated, and the rear one can be shadowed by the ellipsoid surface. Shadowing occurs if the observation angle ν becomes greater than half of the wedge's external angle (see Table 1.3, item 5 about a wedge's RCS for observation angles lying far from normal to the wedge faces and for edge curvature radius $\rho_c \gg \lambda$).

1.5.5 Comparison of Different Approximations of the Blades in JEM and PRM Simulation

All the approximations considered allow us to obtain the JEM and PRM spectra depending on the rotational frequency, blade's angle of attack, and aspect angles of the rotating structure. They all reveal the spectrum asymmetry due to combinations of phase and amplitude modulation.

The approximation with flat plates for single-engine and single-stage JEM using physical optics coincides with those previously described in

[21–26]. Theoretical results of the works [24, 25], verified for the JEM in the field experiments, were justified by their authors for several wavelengths. Theoretical results of the work [23], verified for the one-stage fan in laboratory experiment, were also justified by their authors for specific wavelengths. Our simulation for multistage and multiengine JEM coincides qualitatively with the short-wave experimental results of [29].

To account for the specular point shift in the PRM simulation and the additional modulation it causes, the approximation of blades with twisted plates was proposed [27]. Such additional modulation showed itself more explicitly in our PRM simulation for the blade "ellipsoid-wedge" approximation, accounting for the polarization effects and edge backscattering also.

In the authors' experience, acceptable simulation results can be relatively simple to obtain using blade approximations in two ways:

1. With flat plates for the turbine modulation, taking into account the proposed coefficient of departure from physical optics [absolute value of (1.48)];

2. With the simplest components for the propeller modulation. We rejected approximation of propeller blades with plates due to extreme contrast of scattering angle dependency, this being practically absent for the rounded propeller cross section.

1.6 Radar Quality Indices to Be Simulated

The aim of simulation is not only the study of physical phenomena but also the resulting optimization of radar systems for recognition (Chapters 2–4), detection, and measurement (Chapter 6). Quality indices for these systems must be introduced in advance. We introduce below the recognition quality indices in Section 1.6.1 and the detection and measurement indices in Section 1.6.2.

1.6.1 Quality Indices of Recognition

Alphabet of Objects to Be Simulated. Objects of radar recognition can be targets' classes or types. The alphabets $A_1, A_2, \ldots, A_k, \ldots, A_K$ of these objects and decisions $\hat{A}_1, \hat{A}_2, \ldots, \hat{A}_i, \ldots, \hat{A}_K$, being made in recognition, are assumed to be predetermined.

Conditional Probability Matrix of Decision-Making. In the stationary decision-making process, the conditional probabilities of decision-making $P(\hat{A}_i | A_k) =$

$P_{i|k}$, where $k = 1, 2, \ldots, K$, can be introduced. The probabilities $P_{i|k}$, where $i = k$, are the conditional probabilities of correct decisions. The probabilities $P_{i|k}$, where $i \neq k$, are the conditional probabilities of error decisions. Matrix $\| P_{i|k} \|$ of $K \times K$ dimension is known as a conditional probability matrix of decision-making. It contains conditional probabilities of correct decisions as its diagonal elements, and conditional probabilities of error decisions as its nondiagonal ones. The sum of elements for each row k is equal to unity $\sum\limits_{i=1}^{M} P_{i|k} = 1$.

Conditional Mean Risk (Cost) of Recognition Decision-Making. To compare various decisions, some positive or zero penalties r_{ik} for error $i \neq k$ decisions and zero or negative penalties $-r_{ii} < 0$ (premiums $r_{ii} > 0$) for correct $i = k$ ones can be defined. Matrix $\| r_{ik} \|$ of $K \times K$ dimension is known as a *cost matrix*. By varying the cost matrix, one can describe the performance of each radar system attempting recognition. The simplest cost matrix, which is being used in theoretical consideration, is the negative value of the identity cost matrix (simple cost matrix). A more general diagonal cost matrix (quasi-simple one) with elements

$$r_{ik} = -r_i \text{ if } i = k, \quad \text{and} \quad r_{ik} = 0 \text{ if } i \neq k \tag{1.52}$$

can also be used to define unequal requirements for recognizing target classes or types [31].

The conditional mean risk (mean cost) of decision-making

$$\bar{r}(i) = \sum\limits_{k=1}^{K} r_{ik} P_{i|k} P_k \tag{1.53}$$

is frequently used as a quality index of recognition. Each P_k, $k = 1, 2, \ldots, K$ is here an a priori probability of appearance of the object to be recognized.

The probability of error in recognition, with equiprobable appearance of objects

$$P_{\mathrm{er}} = \frac{1}{K} \sum\limits_{i=1}^{K} (1 - P_{i|i}) = 1 - P_{\mathrm{cor}} \tag{1.54}$$

is a tentative but easy-to-use quality index of recognition. Here, P_{cor} is the probability of correct decision that can also be used as a quality index of recognition.

Entropy of Situations Before and After Recognition. Let us begin again with equiprobable appearance of any object from K possible ones. Ambiguity of such an appearance can be described by the a priori probability $P_k = 1/K$. Entropy of the situation before recognition is defined as $H = \log_2 K = \log_2(1/P_k) = -\log_2 P_k$ and it is used as the logarithmic measure of ambiguity. If a priori probabilities $P_k (k = 1, 2, \ldots, K)$ are not equal to each other, the entropy before recognition diminishes and it is defined as an expectancy of partial entropies

$$H_{br} = -\sum_{k=1}^{K} P_k \log_2 P_k \tag{1.55}$$

Entropy of the situation after recognition can be defined as the expectancy of partial entropies of decisions about the ith object presence provided that the kth object is actually present. Since the coincidence probability of such events is $P_{ik} = P_{i|k}P_k$, the entropy after recognition can be written in the form

$$H_{ar} = -\sum_{i=1}^{K}\sum_{k=1}^{K} P_{ik} \log_2 P_{i|k} \tag{1.56}$$

The information measure of recognition quality (IMRQ) can be defined as a difference between (1.55) and (1.56) in the case of equiprobable appearance of objects to be recognized

$$I = H_{br} - H_{ar} = \log_2 K + \frac{1}{K}\sum_{i=1}^{K}\sum_{k=1}^{K} P_{i|k} \log_2 P_{i|k} \tag{1.57}$$

The coarse information measure of recognition quality can be defined on the coarse hypothesis of equalities $P_{i|i} = P_{cor} = \text{const}$ for all conditional probabilities of correct decisions $i = k$, and similar equalities $P_{i|k} = (1 - P_{cor})/(K - 1) = \text{const}$ for all conditional probabilities of error decisions $i \neq k$. Then,

$$I = I_{coarse} = \log_2 K + P_{cor} \log_2 P_{cor} + (1 - P_{cor})\log_2 \frac{1 - P_{cor}}{K - 1} \tag{1.58}$$

With the growth of K, the first term of (1.58) increases while the two others decrease due to P_{cor} diminishing. Therefore, the optimal value K_{opt} maximizing the coarse IMRQ can be found [32].

Potential Signal-to-Noise Ratio (SNR) of Recognition.   Different definitions of the signal-to-noise energy ratio are frequently used. We will define the potential SNR energy ratio as the ratio of the signal energy at the matched filter output (J = W/Hz) to the spectral density of noise in W/Hz. Then, SNR for the matched processing for a wideband signal echoed from the point scatterer will be replaced by the SNR for a narrowband signal that does not provide resolution of the target elements. This allows the introduction of the signal's energy independent from its modulation, and the estimation of additional losses connected with the target recognition and resolution of its elements.

For instance, if the ratio 15 dB is necessary to achieve the required indices of detection quality, and this ratio has to be increased to 22 dB for the aircraft recognition, then the additional energy needed for recognition is 7 dB. In the case of a wideband signal, the losses on the noncoherent integration of the signals backscattered from the resolved target elements and the gain due to decrease of fluctuations are involved here.

Quality Indices Chosen for Simulation (see also Chapters 2–4).   These are:

- The number K and alphabet of objects to be recognized;
- The recognition error probability in equiprobable appearance;
- The IMRQ (or coarse IMRQ), considering both the number of objects K and the probability of recognition errors.

The quality indices chosen for simulation of various recognition systems depend essentially on the potential SNR. Comparison of various recognition systems is advisable usually in the condition of their potential SNR equality. Scattering simulation permits us to avoid expensive experiments in initial research and development steps for the choice of:

- Alphabets of target classes or types to be recognized;
- Recognition signatures and their combinations;
- Illumination signals (see Chapters 2–3);
- Decision rules (see Chapter 4).

1.6.2 Quality Indices of Detection and Tracking

Quality indices of detection (see Chapter 6), as usual, are conditional probabilities of detection D and of false alarm F. Detection can be considered as a signal recognition of $K = 2$ classes: "target plus interference" and "interference only" with the matrix of conditional probabilities of decision-making $\begin{Vmatrix} D & 1-D \\ F & 1-F \end{Vmatrix}$. Scattering simulation permits taking into account such peculiarities of targets as their RCS distribution in dependence on their type (class), orientation, and also from the bandwidth of illumination signal.

Quality indices of tracking (see Chapter 6) are the degree of measurement accuracy in stationary and transient tracking; the time of target's initial track lock-on; the mean interval between appearances of false trajectories and the mean time of their tracking; and the probability of missing the target.

1.6.3 Choice of Quality Indices

The choice of quality indices depends on the objective of the device. Scattering simulation permits taking into account such peculiarities as the accidental displacements of a target's secondary radiation center (Chapter 6) and fluctuation effects due to reflections from the target body and its rotating parts (Chapter 3) in narrowband illumination, and the effects of complex range profiles and range-angle images (Chapter 2) in wideband illumination. A comparison of various detection and tracking systems is advisable usually in the condition of equality of potential SNRs.

References

[1] Barton D. K., *Modern Radar System Analysis*, Norwood, MA: Artech House, 1988.

[2] Knott E. F., J. F. Shaefer, and M. T. Tuley, *Radar Cross Section*, Second Edition, Norwood, MA: Artech House, 1993.

[3] Ruck, G. T., (ed.), *Radar Cross Section Handbook, Vols. 1 and 2*, New York: Plenum Press, 1970.

[4] Crispin J. W., and K. M. Siegel, (eds.), *Methods of Radar Cross Section Analysis*, New York: Academic Press, 1968.

[5] "Secondary Radiation of Radio Waves." In *Theoretical Foundations of Radar*, Y. D. Shirman (ed.), Moscow: Sovetskoe Radio Publishing House, 1970, pp. 24–83 (in Russian); Berlin: Militär Verlag, 1977, pp. 37–101 (in German).

[6]   "Secondary Radiation and Modulation Effects of Active Radar." In *Handbook: Electronic Systems: Construction Foundations and Theory*, Y. D. Shirman (ed.), Moscow: Makvis Publishing House, 1998, pp. 126–186, Kharkov, Second edition (printing in Russian).

[7]   *Proc. IEEE*, Vol. 53, August 1965 (thematic issue).

[8]   *Proc. IEEE*, Vol. 79, May 1989 (thematic issue).

[9]   Chamberlain, N., E. Walton, and E. Garber, "Radar Target Identification of Aircraft Using Polarization Diverse Features," *IEEE Trans*. AES-27, January 1991, pp. 58–66.

[10]   Jouny I., F. D. Garber, and S. Anhalt, "Classification of Radar Targets Using Synthetic Neural Networks," *IEEE Trans.*, AES-29, April 1993, pp. 336–344.

[11]   Kazakov, E. L., *Space Objects' Radar Recognition by Polarization Signatures*, Odessa Inst. of Control and Management, 1999 (in Russian).

[12]   Shirman, Y. D., et al., "Methods of Radar Recognition and Their Simulation," *Zarubeghnaya Radioelectronika—Uspehi Sovremennoi Radioelectroniki*, No. 11, November 1996, Moscow, pp. 3–63 (in Russian).

[13]   Sukharevsky, O. I., and A. F. Dobrodnyak, "The Scattering by Finite Ideally Conducting Cylinder with Edges Absorbing Coating in Bystatic Case," *Izvestia Vysshih Uchebnyh Zavedeniy, Radiofizika*, Vol. 32, December 1989, pp. 1518–1524.

[14]   Mittra, R., (ed.), *Computer Techniques for Electromagnetics*, Oxford: Pergamon, 1973.

[15]   Solodov, A. V., (ed.), *Engineer Handbook of Space Technology*, Moscow: Voenizdat Publishing House, 1977 (in Russian).

[16]   Kobak, V. O., *Radar Reflectors*, Moscow: Sovetskoe Radio Publishing House, 1975 (in Russian).

[17]   Dobrolensky, Y. P., *Dynamics of Flight in Disturbed Atmosphere*, Moscow: Mashinostroenie Publishing House, 1969 (in Russian).

[18]   *Atmosphere, Handbook*, Leningrad: Hydrometeoizdat Publishing House, 1991 (in Russian).

[19]   Astapenko, P. D., *Aviazionnaya Meteorologiya*, Moscow: Transport Publishing House, 1985 (in Russian).

[20]   Chernyh, M. M., and O. V. Vasilyev, "Coherence Experimental Evaluation of Radar Signal Backscattered by Aerial Target," *Radiotehnika*, February 1999, N2, pp. 75–78 (in Russian).

[21]   Nathanson, F. E., *Radar Design Principles,* New York: McGraw-Hill, 1969, pp. 171–183.

[22]   Bell, M. R., and R. A Grubbs, "JEM Modeling and Measurement for Radar Target Identification," *IEEE Trans.*, AES-29, January 1993, pp. 73–87.

[23]   Tardy, I., et al., "Computational and Experimental Analysis of the Scattering by Rotating Fans," *IEEE Trans.*, AP-44, No. 10, October 1996.

[24]   Piazza, E., "Radar Signals Analysis and Modellization in the Presence of JEM Application to Civilian ATC Radars," *IEEE AES Systems Magazine*, January 1999, pp. 35–40.

[25] Cuomo, S., P. F. Pellegriny, and E. Piazza, "A Model Validation for the "Jet Engine Modulation" Phenomenon," *Electronics Letters*, Vol. 30, No. 24, November 1994, pp. 2073–2074.

[26] Sljusar, N. M., and N. P. Biryukov, "Backscattering Coefficients' Analysis for Airscrew Metallic Blades of Rectangular Form," *Applied Problems of Electrodynamics*, Leningrad: Leningrad Institute of Aviation Instrumentation, 1988, pp. 115–122 (in Russian).

[27] Slyusar, N. M., "Scattering of Electromagnetic Waves by the Blade Model with the Geometric Twist," *Radiotekhnika i Elektronika*, Minsk, 1990, Issue 19, pp. 131–136 (in Russian).

[28] Kravzov, S. V., and S. P. Leshchenko, "Simulation of Electromagnetic Field Scattering on the Airscrew of Aerodynamic Target," *Electromagnetic Waves and Electronic Systems*, Vol. 4, No. 4, 1999, pp. 39–44 (in Russian).

[29] Chernyh, M. M., et al., "Experimental Investigations of the Information Attributes of Coherent Radar Signal," *Radiotekhnika*, No. 3, March 2000, pp. 47–54 (in Russian).

[30] Abramovitz, M., and I. Stegun, (eds.), *Handbook of Mathematical Functions*, Russian edition, Moscow: Nauka Publishing House, 1979.

[31] Shirman, Y. D., "About Some Algorithms of Object Classification by a Set of Features," *Radiotekhnika i Elektronika* 40, July 1995, pp. 1095–1102 (in Russian).

[32] Leshchenko, S. P., "Informational Quality Index of Radar Recognition Systems," *Zarubeghnaja Radioelectronica—Uspehi Sovremennoi Radioelectroniki*, No. 11, November 1996, pp. 64–66 (in Russian).

2

Review and Simulation of Recognition Features (Signatures) for Wideband Illumination

Wideband illumination provides large amounts of recognition information [1–3]. This attracts the attention of radar specialists, and we consider in detail the wideband illumination first in the book. Various definitions of wideband illumination signals and corresponding target recognition features (signatures) are discussed in Section 2.1. Ambiguity functions and processing procedures for the wideband chirp signals with examples of signature simulation are considered in Section 2.2. Simulation of range-polarization and range-frequency signatures for the wideband chirp signal is examined in Section 2.3. We consider peculiarities of the ambiguity function, methods of signature simulation, and processing procedures for wideband stepped frequency (SF) signals in Section 2.4. Targets' two-dimensional (2D) images and their simulation are considered in Section 2.5.

2.1 Definitions and Simulated Signatures for Wideband Signal

A signal can be defined as wideband with respect to its carrier frequency, to its inverse duration, to its possible inverse delay on the antenna aperture, and to an absolute value of its band. The same signal can be wideband for all these different definitions, but not always. In this book the signal is considered wideband if its bandwidth is sufficient for range resolution of aerial target elements. According to the last definition, the following signals are considered here as wideband: linear-frequency-modulated (LFM or chirp) and stepped frequency (frequency manipulated) with contiguous or separated "monofrequency" elements. Use of very short pulses and wideband phase-modulated pulses is also of interest.

Signal signatures of targets used for wideband illumination are:

- Range profiles (RPs) obtained in high range resolution (HRR) radar [1–8];

- Total radar cross sections on the RP base (i.e., the RCS sums of resolved elements) [5–8];

- Range-polarization profiles (RPP) obtained in HRR radar with dual polarization reception [5–8];

- Range-frequency profiles (RFP) obtained in HRR radar by a significant, but not very great increase of "radar-target" contact time [5, 6, 8];

- Two-dimensional images obtained in HRR radar by a considerable increase of the "radar-target" contact time [1, 2, 9].

Information about the target trajectory signatures, including altitude, velocity, etc., is discussed in Section 4.1.3.

2.2 Simulation of Target Range Profiles and RCSs for Wideband Chirp Illumination

We first discuss the simplified and improved method of simulation for the chirp illumination (Section 2.2.1). We then describe the variants of range profile signatures (Section 2.2.2), the simulation of range profiles (Section 2.2.3), and wideband RCS (Section 2.2.4) for models of real targets under various conditions. Comparison of the simulated data with the available experimental results (Section 2.2.5) finishes the discussion.

2.2.1 Simulation Methods for the Chirp Illumination

Simplified Method of Simulation. Let us suppose that (1) the ratio of the LFM pulse bandwidth to the carrier frequency doesn't exceed 0.05–0.1, (2) the illumination LFM signal has a rectangular or Gaussian envelope, and

(3) matched signal processing is carried out. Expression $U(t)$ for the processed signal entered into (1.20) is proportional then to the Woodward ambiguity function for the rectangular envelope [10–12]

$$\rho(\tau, F) = \begin{cases} \dfrac{\sin[\pi(n\tau/\tau_p + F\tau_p(1 - |\tau|/\tau_p)]}{\pi(n\tau/\tau_p + F\tau_p)}, & |\tau| \leq \tau_p \\ 0 & |\tau| > \tau_p \end{cases} \quad (2.1)$$

or for the Gaussian one

$$\rho(\tau, F) = \exp\left(-\frac{\pi}{2}\left[\frac{1 + n^2}{\tau_p^2}\tau^2 + 2n\tau F + \tau_p^2 F^2\right]\right) \quad (2.2)$$

Here, τ, F are the mismatches in time delay and frequency, τ_p is the pulse duration, measured for the Gaussian pulse at the level $e^{-\pi/4} \approx 0.46$, $n = \tau_p \Delta f$ is the bandwidth-duration product, Δf is the frequency deviation measured for the Gaussian pulse at the level $e^{-\pi/4}$.

The sidelobes of the compressed pulse are absent in case (2.2) and are significant in case (2.1). To diminish the sidelobe level of the compressed LFM pulse with rectangular envelope, Hamming weighting is often used:

$$U_1(t) = bU(t) + aU(t - 1/\Delta f) + bU(t - 2/\Delta f) \quad (2.3)$$

where $a = 0.5$ and $b = 0.25$, for instance. To exclude such detail from simulation, the Gaussian pulsed illumination model can be used.

As was noted in Section 1.3.4, the signal backscattering and filtration are *linear operations*. Real illumination with a modulated signal [Figure 2.1(a)]

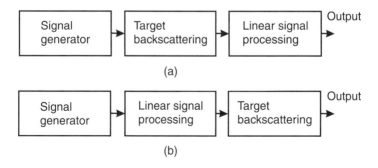

(a)

(b)

Figure 2.1 Clarification of the change in order of linear operations in simulation.

can be replaced therefore by illumination with a short "monofrequency" pulse obtained after processing [Figure 2.1(b)] with an envelope of the form (2.1) or (2.2). The approach for time-extended illuminating signals, when Doppler shift is essential, will be developed in Sections 2.4.2 and 2.4.3.

Possibility of Developing the Simplified Simulation Variant for the Chirp Illumination. If the ratio of LFM pulse bandwidth to the carrier exceeds 0.05–0.1, one can consider this LFM pulse as the superposition of partial LFM pulses of a shorter duration and lower frequency deviation. This is especially simple for pulses with rectangular envelope when the contiguous partial LFM pulses $i = 1, 2, \ldots, m$ are rectangular and of equal duration and frequency deviation but of various carrier frequencies. After delay alignment and superposition of backscattered and compressed partial signals of the (2.1) type with $\exp[j2\pi(f_i - f_0)t]$ multipliers, one can obtain the desirable simulated compressed signal. Then the simulation program has to schedule the evaluation of the ratio of given pulse bandwidth to the carrier and the described correction of the simplified simulation. Our development of simulation based on the data presented in Table 1.3 accounts for the dependence of RCS variations on frequency. Unfortunately, additional phase-delay variations are not accounted for. This can diminish the accuracy of signal simulation for very large ratios of the bandwidth to the carrier.

2.2.2 Variants of Signatures on the Basis of Range Profiles

The RP of a target is formed at the amplitude detector output, provided that target elements are resolved in range. The larger the target radial extent, the more extended is its RP. Moreover, some other target structure feathers, such as positions of rotational modulation sources, are mapped on the RPs. The phases of a single RP itself are usually noninformative because of its excessive dependence on the target aspect. The use of single RPs, therefore, will be connected below to *amplitude detection*. Replacing the amplitude detector by two phase detectors, we obtain the quadrature RP, which can be used for coherent integration from pulse-to-pulse and Doppler frequency analysis. The use of phase detection is considered in this book, however, only for such special variants of RP use.

Sampled and Normalized RPs. Digital processing often requires the RP to be sampled into a sequence $X_1, X_2, \ldots, X_M$ after amplitude detection (i.e., with phase data excluded because of their instability). The cited sequence may be briefly described by a vector-column $\mathbf{X} = \| X_1, X_2, \ldots, X_M \|^{\mathrm{T}}$.

Each sampled RP is normalized to the unit power, so that $\mathbf{X}_n = \mathbf{X}/|\mathbf{X}|$, where $|\mathbf{X}| = \sqrt{\sum_{m=1}^{M} X_m^2}$. If $|\mathbf{X}| = 1$, then $\mathbf{X}_n = \mathbf{X}$. The RP scale is unified due to normalization. Normalization allows us to extract the information about the structure using it independently from the amplitude information. The latter is influenced to a great extent by a series of factors: output transmitter power, receiver sensitivity, and radio wave propagation conditions. Therefore, we must use it with the lower weight for decision-making.

Correlation processing of the RP consists frequently in evaluation of correlation sum

$$Z_\Sigma = \sum_m Y_m X_m = \mathbf{Y}^T \mathbf{X} \qquad (2.4)$$

where $\mathbf{Y} = \| Y_1 \quad Y_2 \quad \ldots \quad Y_m \quad \ldots \quad Y_M \|^T$ is the sample vector of the received RP amplitudes $Y_m = |\dot{Y}_m|$ and $\mathbf{X} = \| X_1, X_2 \ldots X_m, \ldots X_M \|^T$ is the sample vector of reference RP amplitudes. The reference vectors $\mathbf{X} = \mathbf{X}_k$ differ from one another for various targets, their orientations, and echo delays. To consider the unknown target's orientation in recognition decision-making, a set of standard vectors (standard RPs) $\mathbf{X}_{\gamma|k}$ can be introduced in the ambiguous aspect sector. To account also for unknown echo delay (index μ), one has to evaluate the maximum correlation sum

$$Z_{\Sigma \max} = \max_\gamma \max_\mu \sum_m |\dot{Y}_m| X_{(m-\mu)\gamma|k}$$

The grounding and developing of this procedure are discussed in Section 4.1.4.

Standard RPs $\gamma = 1, 2, \ldots, \Gamma$ are the normalized reference RPs, obtained on the basis of experiment or simulation to be used in correlation processing. Each of $T \geq \Gamma$ teaching RPs can be considered as a standard one. But the number Γ of standard RPs then increases to the greater number T of teaching RPs, slowing the decision-making.

Simple Procedure of Standard RPs Formation. To form $\Gamma < T$ of standard RPs, the following heuristic recursive procedure was used in our 1985–1987 experiment (see Section 2.2.5), as well as in subsequent simulation for signal bands narrower than 100 MHz. The training set of each target class included all training RPs of various target types. At its preliminary step, the correlation of each training RP with rectangular "profiles" of different extent was found. The extent of the rectangular "profile" that gave maximum correlation with the teaching RP was then accepted as the conditional extent of this RP. Then, all T teaching RPs are ranked and partitioned into Γ clusters according to their conditional extent. The RPs of each cluster are centered using correlation procedure, and the mean RPs are found. Such "mean" RPs are considered as the standard RPs of clusters at the preliminary step.

At the first and several later steps the correlation of each of T teaching RPs with each of Γ standards of the previous step is found. The teaching RP is then directed into the cluster whose standard RP has provided maximum correlation. The "mean" RPs of the Γ new clusters are considered then as the new RP standards, and the procedure is either repeated or stopped [8]. The standard RPs obtained using the simplest procedure of their formation will be named the simple standard RPs.

Individualized Procedure of Standard RPs Formation. This procedure is analogous to that described just above, but it envisages formation of the standard RPs separately for various target types. A set of standard RPs of target class consists then of the standard RPs of target types included in the target class. The standard RPs obtained using the individualized procedure of their formation will be named the individualized standard RPs. As it will be shown in Section 4.1.4, the individualized procedure of standard RPs formation provides better quality of recognition than the simplified one. But it usually requires the use of a greater overall number of standard RPs, as well as of processing channels. It also can be less robust to the deformation of RPs by the suspension of fuel tanks and additional armament, and also to the appearance of new types of targets. Robustness can be extended by increasing the total number of RPs.

Conditional Probability Density Function (cpdf) of RP and Its Logarithm (lcpdf). Each RP sample is a random variable, whose distribution can be described by a one-dimensional (1D) conditional cpdf for the given target type or class, its range, and aspect sector. An approximation of the samples' independence simplifies consideration. The multidimensional cpdf is then equal to the product of the 1D samples' cpdf. The lcpdf is equal to the sum of the 1D lcpdfs of samples. Thus,

$$\text{cpdf}(\mathbf{Y}) = \prod_m \text{cpdf}(Y_m), \qquad \text{lcpdf}(\mathbf{Y}) = \sum_m \text{lcpdf}(Y_m) \qquad (2.5)$$

All 1D cpdfs and lcpdfs can be obtained from experiment or simulation (see below). Multidimensional cpdfs and lcpdfs can be used instead of correlation sums in recognition decision-making [13], see also Section 4.1.

2.2.3 Simulation of the Target RPs

Examples of simulated recurrent RPs superimposed on the screen of the amplitude indicator for Tu-16- and Mig-21-type aircraft and a ALCM-type missile are shown in Figure 2.2 and Figure 2.3. Examples of Figure 2.2 are given for the Gaussian chirped illumination pulses of about 80 and 200 MHz. It can be seen that for the 200-MHz deviation the RP is presented in more detail; however, the 80-MHz deviation provides the possibility of the target's class recognition. It can be seen that targets of various classes can be distinguished simply by the extent of their RPs. Examples of Figure 2.3 are given for the Gaussian chirped illumination pulses of about 80-MHz frequency deviation and for two aspects (course-aspects) differing by 0.1° (from top to bottom). For a signal bandwidth narrower than 100 MHz, small orientation variations of the target notably influence the shape of the large RPs (Tu-16 type in Figure 2.3). This influence is caused by scatterer interference within a range resolution cell that is wider than 1.5m. For the wider signal bandwidth (Figure 2.2) the RPs become more detailed and the influence of aspect variation decreases. The chosen wavelength of 11.5 cm corresponds to the portion from 2500 to 2700 MHz of radar S-band. The lower RPs of Figure 2.2 are calculated under the hypothetical assumption

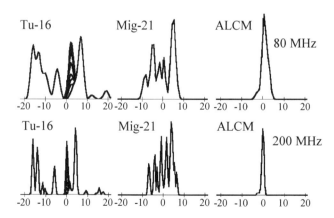

Figure 2.2 Simulated superimposed range profiles of the Tu-16- and Mig-21-type aircraft and the ALCM-type missile illuminated by the chirp signals with frequency deviations of 80 and 200 MHz.

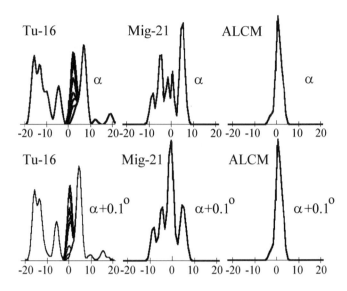

Figure 2.3 Simulated superimposed range profiles of the Tu-16- and Mig-21-type aircraft and the ALCM-type missile illuminated by the chirp signals with frequency deviation of 80 MHz for the aspect change of 0.1°.

of using the entire portion (i.e., of having the 200-MHz bandwidth). At the chosen wavelength of 11.5 cm, a small amplitude blurring of superimposed RPs takes place at the positions of JEM sources.

The simulated RPs of Tu-16-type aircraft are shown in Figure 2.4 for a wavelength of 5 cm, at which the turbine modulation manifests itself more explicitly than in the previous case. The successive RPs are given for illumination with Gaussian chirped pulses with 5, 60, and 160 MHz frequency deviation (from top to bottom). As before, it can be seen that range resolution and profile specification increase as the bandwidth widens. By the successive time realizations RP_1, RP_2, RP_3, . . . , RP_{1001}, RP_{1002}, RP_{1003} (from the left to the right with time gap more than 500 ms), the time dynamics of RPs are traceable. For the small deviation of 5 MHz, almost the whole signal (a kind of RP!) fluctuates quickly at λ = 5 cm due to orientation change and turbine modulation. For the deviations of 60 and 160 MHz, only a part of RP fluctuates quickly at λ = 5 cm due to turbine modulation, and the rest of RP changes much more slowly.

Dynamics of the RPs (RP_1, RP_2, RP_3, . . . , RP_{1001}, RP_{1002}, RP_{1003}) for B-52-type aircraft in flight are given in Figure 2.5 for the 160-MHz deviation signal and two wavelengths λ = 5 cm [Figure 2.5(a)] and λ = 15 cm [Figure 2.5(b)]. RPs of B-52-type aircraft have a somewhat larger

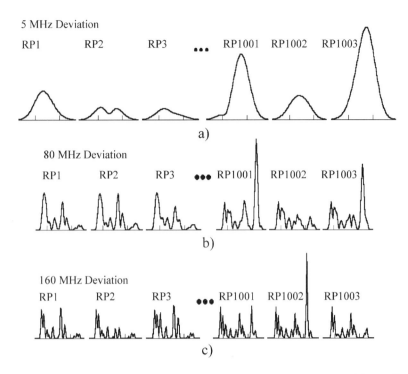

Figure 2.4 Simulated successive range profiles RP$_1$, RP$_2$, RP$_3$, ..., RP$_{1001}$, RP$_{1002}$, RP$_{1003}$ backscattered from Tu-16-type aircraft in flight for chirp illumination with frequency deviations of (a) 5 MHz, (b) 60 MHz, and (c) 160 MHz. Wavelength λ is 5 cm; PRF is about 1 kHz.

extent than those of Tu-16-type aircraft. As in the previous case, only parts of the RPs fluctuate quickly due to turbine modulation on 5-cm wavelength. The turbine modulation is decreased at the 15-cm wavelength.

In the case of a helicopter all the RPs fluctuate quickly because its rotors have sizes comparable with the helicopter itself and are placed in horizontal and vertical planes. Figure 2.6 presents these RPs simulated dynamically for 160-MHz frequency deviation and a wavelength λ = 15 cm.

Changeability and recurrence of RPs depend on the type of target, its orientation, and the signal bandwidth. The influence of rotational modulation on the RPs was discussed in connection with Figures 2.4 and 2.5.

Changeability due to orientation shift depends on the signal bandwidth. The degree of changeability can be described by the coefficient ρ of correlation between RP pairs for a given aspect shift $\Delta\alpha$. Such a correlation coefficient versus signal's bandwidth $\rho = \rho(B|\Delta\alpha)$ is shown in Figure 2.7 for the large-sized target and constant aspect shifts of 0.2°, 1°, 10°. The

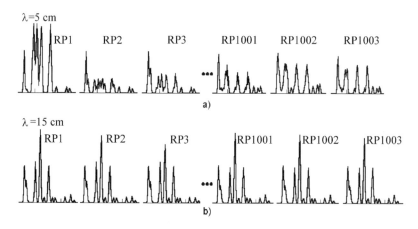

Figure 2.5 Simulated successive range profiles RP$_1$, RP$_2$, RP$_3$, ..., RP$_{1001}$, RP$_{1002}$, RP$_{1003}$ backscattered from B-52-type aircraft in flight for chirp illumination with 160-MHz frequency deviation and wavelengths (a) λ = 5 cm and (b) λ = 15 cm. PRF is about 1 kHz.

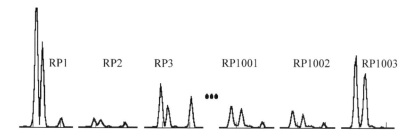

Figure 2.6 Simulated successive range profiles RP$_1$, RP$_2$, RP$_3$, ..., RP$_{1001}$, RP$_{1002}$, RP$_{1003}$ backscattered from AH-64 helicopter in flight for chirp illumination with 160-MHz frequency deviation and λ = 15 cm. PRF is about 1 kHz.

correlation is large for small bandwidths because the RPs slightly depend on the target aspect (also carrying little information). RPs become more informative as the bandwidth grows, but correlation decreases due to the fluctuations of partly resolved target elements. The correlation coefficient reaches its minimum at 40 to 60 MHz bandwidths for $\Delta\alpha$ = 0.2° to 1°. It increases then because fewer bright elements fall into the range resolution cell. Recurrence of RPs in $\Delta\alpha$ shift therefore take place. The correlation gradually decreases with increasing bandwidth for large aspect shifts $\Delta\alpha$, and the RPs become more individual and informative for each target's orientation.

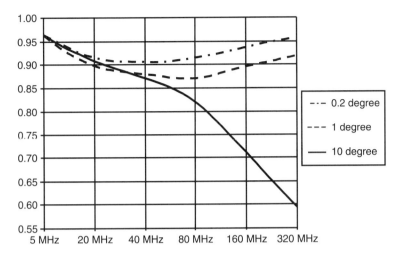

Figure 2.7 Correlation of the simulated range profile couples versus signal bandwidth for aspect shifts of 0.2°, 1°, and 10°.

An example of simulated cpdf of RP samples [13] is shown in Figure 2.8 for a given large-sized target in a noise background of low level. A set l = 1–128 of its one-dimensional cpdfs $p(y_l)$ is depicted. It is interesting that simulated cpdfs are notably different from the Gaussian. When the RP sample is a superposition of signal and noise, its cpdf is a convolution of cpdf of the RP sample without noise and the pdf of noise.

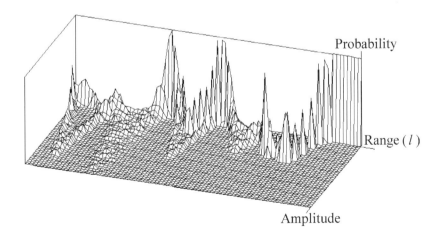

Figure 2.8 Simulated set of cpdf for the range profile samples l = 1–128 of a large-sized target without noise.

2.2.4   Simulation of the Target RCS for Wideband Illumination

As was stated in Section 1.1.1 for wideband illumination of targets, the sum of mean RCSs of its resolved elements is considered as its mean RCS. The RP samples spaced by an interval $\Delta t = 1/\Delta f$ can be considered as the resolved target elements. The known algorithm of experimental RCS evaluation (Section 3.2) must then be applied to individual RP samples. The partial RCSs obtained in such a manner must be summed. Fluctuations of RCS in wideband illumination are essentially diminished, but not completely removed.

Figure 2.9 shows the normalized standard deviation of the simulated RCS estimate versus the number N of independent "radar-target" contacts. Normalization has been provided relative to maximum RCS variance in narrowband illumination. Three solid lines correspond to the RCS of large-sized (1), of medium-sized (2), and of small-sized (3) targets illuminated by the chirped pulse of 80-MHz deviation. The shaded zone between the dashed lines includes such RCS dependencies for various targets illuminated by the narrowband signal. The RCS fluctuations for the wideband illumination are decreased with respect to those of narrowband illumination because of additional averaging by range, so that a smaller number N of "radar-target" contacts is required to estimate the RCS with a given accuracy. Bandwidth widening beyond 80 to 100 MHz also diminishes the fluctuations.

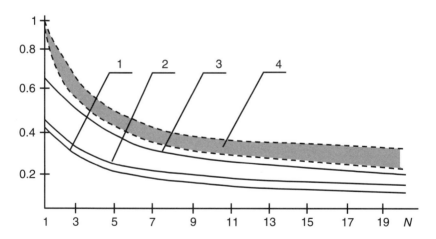

Figure 2.9 Normalized standard deviations of the simulated RCS estimate versus the number of independent radar-target contacts by simulation results: (1) for large-sized (solid line 1), medium-sized (solid line 2), and small-sized (solid line 3) targets with chirp illumination of 80-MHz deviation; (2) for targets of various dimension with narrowband illumination (shaded zone between dashed lines).

Aspect dependencies of the target's RCS can be used not only for narrowband but also for wideband illumination. Figure 2.10 shows these simulated dependencies after averaging for Tu-16- and Mig-21-type aircraft and ALCM-type missiles.

2.2.5 Comparison of Simulated and Experimental Data

Some Experiments of the 1950s–1960s.   After the reinvention of a pulse compression method unknown at that time in the USSR (1956, [14]), the full-scale radar models using this method were tested near the city of Kharkov:

1. The 1959 "Filtratsiya" model with an aircraft detection range of about 200 km. It worked in the high-frequency part of VHF wave band and was built on the P-12 surveillance radar base. It was characterized by the generation of 5 MHz × 6 μs illumination chirp pulses. It used a compression filter consisting of two delay lines with planar configuration and distributed capacitance coupling [3];

2. The 1963–1964 "Okno" model with an aircraft detection range of about 100 to 150 km. It worked in the high-frequency part of the S-band and was built on the PRV-10 heightfinder base. It was characterized by the generation of 70 MHz × 2 μs illumination chirp pulses. It used a compression filter consisting of a rolled cable of 400m length with 12 oscillatory circuits between the terminals and used a wideband display. The experiment confirmed the possibility of RP use.

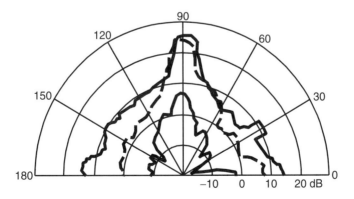

Figure 2.10 Simulated dependencies of RCS versus aspect angle for Tu-16- and Mig-21-type aircraft and an ALCM-type missile for the wideband chirp illumination of 80-MHz deviation after averaging.

Figure 2.11 shows the RPs of An-10, Li-2, and Su-9 aircraft together with their outlines. Target radial dimensions were evaluated by these RPs with accuracy of a few meters, although the potential range resolution was not achieved because of unaccounted-for FM nonlinearity and other hardware limitations [3, 4].

In 1967 the experiment of Bromley and Callen [15] was published on the base of illumination chirp pulses 150 MHz $\times$ 1 μs and a compression filter on the basis of waveguide of 183m length.

Tests analogous to the "Okno" experiments with wideband signals were repeated afterwards in the USSR with the assistance of industrial enterprises in 1980 and 1985–1987.

The 1980 Experiment at the Heightfinder.   Here, the LFM signal of about 50-MHz bandwidth formed more precisely than in 1963–1964 was used in 1980 at a newer radar heightfinder. It worked in the high-frequency part of S-band.

Combined heterodyne-filter signal processing proposed in Kharkov in the 1960s was used in this case. The LFM heterodyne voltage was acting on the mixer in a range gate whose width equaled the sum of the pulse duration in the range units and radial target dimension. Frequency deviation of the pulse scattered from bright points was preserved and decreased due to the scheduled small difference of transmitter and heterodyne LFM slopes. Heterodyning simplified the matched processing, allowing use of ultrasonic compression filters. Besides, unequal frequency displacements of signal elements by the heterodyning cause unequal time displacements at the compression filter output. Target RP is observed therefore in the stretched time scale. This technique has become known in Western literature as "stretch processing" [16].

Various targets' RPs, including the successive ones (without integration), were observed by the described radar. The slow ramp voltage was

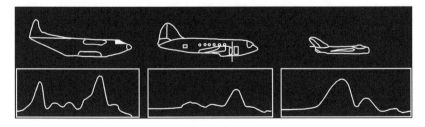

Figure 2.11   The first experimental range profiles of An-10, Li-2, and Su-9 aircraft in chirp illumination (1963–1964) with their outlines.

superimposed in the last case on the output RP. Figure 2.12 shows that the RPs of an incoming target were not being distinctly changed during the burst time, if only rotational modulation could be neglected; so, aerial target RP noncoherent integration is possible. On a level with aircraft RPs, those of a helicopter were obtained too. Due to their blade rotation, even the two successive RPs were not alike. Range-polarization profiles were also obtained (see Section 2.3).

1980s Experiments with a Three-Coordinate Radar. Digital records of range profiles with and without noncoherent integration were obtained in 1985–1987 at a 3D surveillance radar operated in the high-frequency part of S-band with an aircraft detection range of more than 250 km, a scan period of 10 sec, and narrowband illumination. Wideband LFM illumination with deviation of about 75 MHz was used in a narrow azimuth sector ±1.5° within a chosen antenna rotation cycle as the complementary recognition mode implemented after target detection. Wideband LFM pulses were preceded by more narrowband LFM pulses (5 MHz) to realize robust heterodyne-filter processing. The complementary LFM information was used for precise timing of the gate of the LFM heterodyne, accounting for the target radial

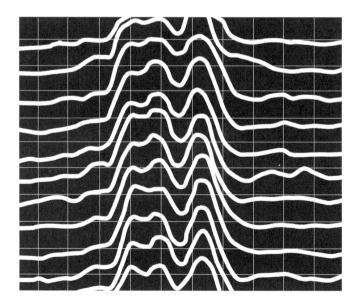

Figure 2.12 Successive experimental range profiles for chirp illumination signal with bandwidth of about 50 MHz observed by superimposing the slow ramp voltage onto them.

velocity. The RPs of each burst were sampled, integrated, and recorded. Single (nonintegrated) RPs could also be recorded.

Automatic target class recognition was realized by means of the coarse evaluation of the target 10-degree aspect sector and correlation processing of received RPs using simple standard RP variants for this aspect sector together with the result of wideband RCS evaluation. The pictogram of target class then followed the target blip on the plan-position indicator (PPI) display until it disappeared. Figure 2.13 presents some experimental RPs of Tu-16, Tu-134, Mig-21, Sy-27 aircraft and a meteorological balloon with a reflector. The experimental dependencies of the wideband RCS via their aspect angle for the Tu-16 and Mig-21 aircraft and balloon with a reflector are shown in Figure 2.14. They are similar to the simulated ones (Figure 2.10).

Some Experiments of the 1990s. In their 1993 paper Hudson and Psaltis reported the acquisition and use of a large volume of RPs with individualized standard RPs, achieved by illumination of many types of targets with a 300-MHz bandwidth chirp signal (see Section 4.2 and [17]). Chirp signals were also used in European radar: in the "Ramses" mobile radar of 200-MHz bandwidth, and in the "TIRA" stationary radar of 800-MHz bandwidth, the latter in Ku-Band. Simultaneously, the stepped frequency signals of 200-MHz bandwidth are studied in the "MPR" mobile radar in X-Band and of 400-MHz bandwidth in the "Byson" stationary radar in S-Band [18]. A chirp signal of 400-MHz bandwidth was also realized in the Chinese experimental radar [19]. The absence of detailed publications did not allow us to compare our simulated results with the above-mentioned experimental ones.

Comparison of Simulated and Experimental Results. The lack of precise target orientation in an experiment obstructs to a certain degree comparison of simulated and experimental data. The comparison was conducted therefore in the statistical sense. The simulated simple standard RPs were compared with those obtained in the 1985–1987 experiments. The correlation coefficient of simulated and experimental simple standard RPs of aircraft happened to fall within 0.88 to 0.97 limits for three simple standards in a 10-degree aspect sector. It would have been even higher if the number of standards had been increased.

The comparison was done not only for averaged but for single-pulse profiles. The simulated RPs of Tu-16- and Mig-21-type aircraft and the ALCM-type missile (Figure 2.3) are close to those obtained in the experiment

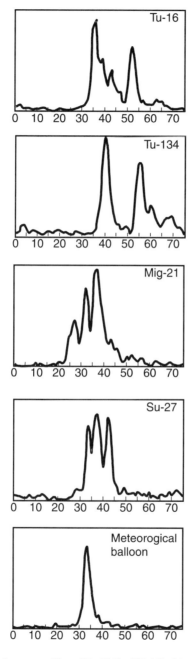

Figure 2.13   Experimental range profiles of Tu-16, Tu-134, Mig-21, and Su-27 aircraft and the balloon with a reflector for chirp illuminating signal with frequency deviation of 75 MHz.

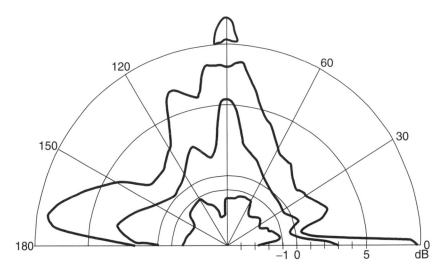

Figure 2.14 Experimental wideband RCS (after averaging) versus aspect angle for Tu-16 and Mig-21 aircraft and balloon with a reflector for chirp illumination with 75-MHz deviation.

using a balloon with a reflector instead of an ALCM-type missile (Figure 2.13). In any case, the deviation between simulated and experimental RPs does not exceed the simulated RP deviation due to the target aspect change of fractions of degree (Figures 2.3 and 2.13). Additional comparisons are carried out for the narrowband illumination RCS in Section 3.2 and for rotational modulation spectra in Section 3.3.

2.3 Range-Polarization and Range-Frequency Signatures Simulation for the Chirp Illumination

Section 2.3 considers the range-polarization (Section 2.3.1) and range-frequency signatures (Section 2.3.2) together with the results of their simulation and some experimental results. The consideration is carried out in the context of chirp illumination. The results obtained for range-polarization signatures can be applied also to the case of stepped-frequency illumination. The results received for range-frequency signatures have some specifics.

2.3.1 Range-Polarization Signatures and Their Simulation

Range-Polarization Signatures. The target consists of both irregular and smooth elements. Its range profiles and wideband RCSs depend therefore on the polarization of transmitting and receiving antennas if the wavelength is not very short. The HRR radar can then obtain complementary information about the target due to complete polarization transmission-reception. An attempt to use such information was described in [7], where the evaluation of polarization ellipse parameters for each resolved target element was suggested. Such a procedure has a definite drawback, that is, the necessity of precise parameter measurement before integration of signal energy. Cooperative information extraction from all the RPP elements is more admissible in cases of low energy. Two kinds of such RPP are possible: noncoherent RPP and coherent RPP.

Noncoherent RPP $\mathbf{X}$ includes the M two-amplitude signal samples X'_m, X''_m ($m = 1, 2, \ldots, M$) for orthogonal polarization instead of the M one-amplitude ones X_m in the considered RP

$$\mathbf{X} = \left\| \mathbf{X}_1 \quad \mathbf{X}_2 \quad \ldots \quad \mathbf{X}_m \quad \ldots \quad \mathbf{X}_M \right\|^{\mathrm{T}}$$

where

$$\mathbf{X}_m = \left\| X'_m \quad X''_m \right\|^{\mathrm{T}} \tag{2.6}$$

It allows us to accumulate energy using the set of noncoherent received signal-plus-noise samples

$$\left\| \mathbf{Y}_1 \quad \mathbf{Y}_2 \quad \ldots \quad \mathbf{Y}_m \quad \ldots \quad \mathbf{Y}_M \right\|^{\mathrm{T}} = \mathbf{Y}$$

where

$$\mathbf{Y}_m = \left\| Y'_m \quad Y''_m \right\|^{\mathrm{T}}$$

using a correlation procedure

$$Z_{\mathrm{p\,noncoh}} = \left| \mathbf{Y}^{\mathrm{T}} \mathbf{X} \right| \tag{2.7}$$

Coherent RPP $\dot{\mathbf{X}}$ accounts for two amplitude samples X'_m, X''_m of orthogonal polarization and the phase difference β_i between the corresponding sampled sinusoids:

$$\dot{\mathbf{X}} = \left\| \dot{\mathbf{X}}_1 \quad \dot{\mathbf{X}}_2 \quad \ldots \quad \dot{\mathbf{X}}_m \quad \ldots \quad \dot{\mathbf{X}}_M \right\|^{\mathrm{T}}$$

where

$$\dot{\mathbf{X}}_m = \left\| X'_m \quad X''_m \exp(j\beta_m) \right\|^{\mathrm{T}}$$

For a fixed transmitter polarization it utilizes the same information as the set of polarization ellipses [7], allowing, however, energy integration using the correlation procedure

$$Z_{\text{p coh}} = \left| \dot{\mathbf{Y}}^{*\mathrm{T}} \dot{\mathbf{X}} \right| \tag{2.8}$$

over the set of coherent received samples

$$\left\| \dot{\mathbf{Y}}_1 \quad \dot{\mathbf{Y}}_2 \quad \ldots \quad \dot{\mathbf{Y}}_m \quad \ldots \quad \dot{\mathbf{Y}}_M \right\|^{\mathrm{T}} = \dot{\mathbf{Y}}$$

where

$$\dot{\mathbf{Y}}_m = \left\| Y'_m \quad Y''_m \arg(\dot{Y}''_m / \dot{Y}'_m) \right\|^{\mathrm{T}}$$

The Y'_m, Y''_m are amplitudes and $\dot{Y}'_m$, $\dot{Y}''_m$ are complex amplitudes of the polarization channel output voltages.

Independent signal radiation at two polarizations demands an appreciable hardware volume increase and doubled energy consumption. With an illumination signal of fixed polarization, a pair of channels, mismatched identically relative to polarization of the illumination signal, can also be used as a pair of reception channels with orthogonal polarizations:

- A pair of reception channels with circular polarization and opposite rotation directions in case of linear polarization of the illumination signal;

- A pair of reception channels with orthogonal linear polarization in the cases of illumination signals with (1) circular polarization and (2) linear polarization mismatched at $\pm 45°$ relative to polarization of reception channels.

Some kinds of RPP normalization are possible. One consists in normalization of both RPP components taken as a whole. Together with the normalized RPP, the summed wideband RCS at both orthogonal polarization can then be used as one of the RPP signatures. A correspondence of the described RPP algorithms and the algorithm based on evaluation and comparison of the polarization matrix parameters will be considered in Section 3.2.2.

RPP Signature Simulation. Figure 2.15 shows two pairs of Tu-16- and Mig-21-type aircraft RPPs simulated for an S-band illumination chirp signal with 50-MHz bandwidth at vertical polarization and for reception of a backscattered signal with two antennas' channels of matched (vertical) and cross (horizontal) polarizations.

Figure 2.16 shows the pairs of the Tu-16- and Mig-21-type aircraft RPPs simulated for the illumination signal having circular polarization, received by two antenna channels of vertical and horizontal polarizations. The system operates in L-band to increase the polarization effects and has a bandwidth of 160 MHz to prevent interference between the target elements. Figure 2.17 shows the phase difference for the two RP pairs of Figure 2.16,

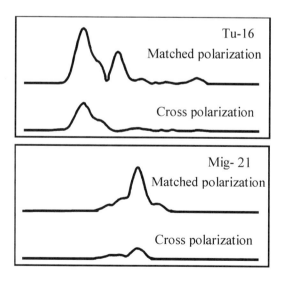

Figure 2.15 Simulated Mig-21- and Tu-16-type aircraft noncoherent range-polarization profiles for illumination with S-band chirp signal of 50-MHz bandwidth of vertical polarization and reception of backscattered signal by two antennas' channels of vertical and horizontal polarization.

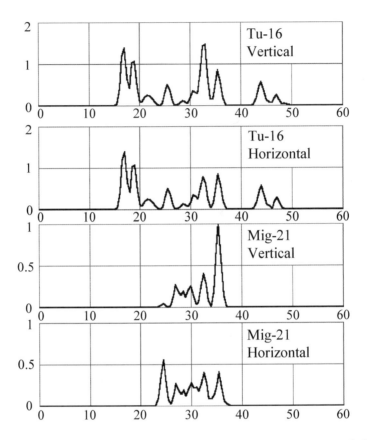

Figure 2.16 Simulated Mig-21- and Tu-16-type aircraft noncoherent range-polarization profiles for L-band illumination signal of 160-MHz bandwidth having circular polarization and for reception by two antennas' channels of vertical and horizontal polarization.

so that the RP pairs of Figure 2.16 together with Figure 2.17 give the coherent RPPs.

The increase in information due to polarization consideration is not very large in S-band. It is greater for the longer waves at L- and UHF-band.

RPP experimental observation was carried out in 1980 for a chirp illumination signal in S-band having a bandwidth of 50 MHz and complete polarization transmission. The reception was at the chosen polarization and of amplitude only. The example (Figure 2.18) corresponds to a linear polarization transmission and reception of matched and crosspolarizations for signals backscattered from Su-15, Yak-28p, and Tu-16 aircraft. The experimental RPP of the Tu-16 aircraft (Figure 2.18) is close to the simulated RPP (Figure

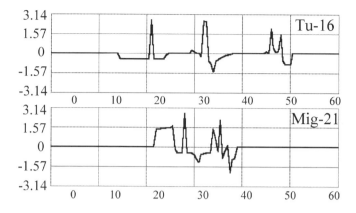

Figure 2.17 Simulated phase differences for the noncoherent range-polarization profiles of Figure 2.16.

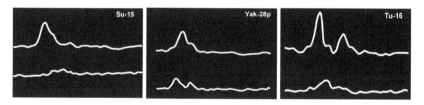

Figure 2.18 Experimental noncoherent range-polarization profiles obtained for chirp illumination in S-band with bandwidth of about 50 MHz for Su-15, Yak-28p, and Tu-16 aircraft.

2.15). The experimental RPPs of the Su-15 and Yak-28p aircraft (Figure 2.18) are close to the simulated one of Mig-21 aircraft corresponding to the same target class.

2.3.2 Range-Frequency Signatures and Their Simulation

Range-Frequency Signatures. Let us consider the coherent pulse burst that is backscattered from a target. The burst consists of N chirp pulses having the duration τ_p, frequency deviation Δf, and repetition period T. A real target is extended in radial and transverse directions, implements in whole, radial, and transverse motions, and has rotating parts. The target's radial extent leads to the RP formation. Its radial movement can lead to the range displacement δr of the RP and changes the range resolution cells where the RP is formed. For the doppler frequency $F_D = 10$ kHz, $\Delta f = 100$ MHz,

τ_p = 30 μs, the displacement δr during a single pulse processing time is less than the range resolution cell's radial extent Δr:

$$\delta r = \frac{c}{2}\tau_\mathrm{p}\frac{F_\mathrm{D}}{\Delta f} = \frac{3 \cdot 10^8 \cdot 3 \cdot 10^{-6} \cdot 10^4}{2 \cdot 10^8} = 0.045\mathrm{m} < \Delta r = \frac{c}{2}\frac{1}{\Delta f} = 1.5\mathrm{m}$$

Pulse displacements for repetition intervals T can be taken into account in the process of coherently integrating the burst's energy.

Let us consider now the influence of a target transverse extent on the coherent integration of energy for the N chirp pulse burst in the case of a target transverse motion or rotation. The increments of radial velocities δv_r and doppler frequencies $2\delta v_r/\lambda$ of the target elements then emerge. Let us consider here the case when the burst duration NT is not very great, so that $1/NT \gg 2\left|\delta v_r\right|_{\max}/\lambda$ (the case of a greater burst duration NT will be considered in Section 2.5). The transverse elements of target body are not yet resolved, but the rotational modulation frequencies can already be resolved in doppler frequency. This leads to the formation of a range frequency profile (RFP) containing information about the location of rotating parts. To obtain the RFP, one must take the discrete fourier transform (DFT) in each range resolution cell over the burst duration.

RFP Simulation.   The example of a simulated An-26 aircraft RFP is shown in Figure 2.19. The vertical axis of Figure 2.19 is range (meters), and the horizontal axis is frequency (kilohertz). On a level with nose N and tail parts T, two propellers P are observed, one of them shadowed for half the record [13].

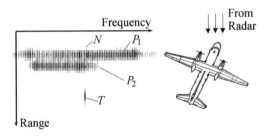

Figure 2.19   Example of the target's range-frequency profile for chirp illumination in C-band with the bandwidth of 160 MHz for An-26 aircraft with the 20-degree aspect angle. Shown are range axis (vertical, m), frequency (horizontal, Hz), target's nose *N*, tail *T*, and two propellers *P*, one of which is shadowed for half the record.

2.4 Target Range Profiles for Wideband SF Illumination

SF illumination signals may be contiguous or separated as is explained in the Gabor diagrams (Figure 2.20). The Gabor diagram presents the signal as a superposition of its time-frequency portions with $\delta f \delta t = 1$ products. Contiguous SF signals similar to the chirp ones are well known. Separated SF signals now attract a special interest. They can provide a wideband mode implementation in existing radar with pulse-by-pulse frequency agility. Separated SF signals can be used both with moderate and very large bandwidth-duration products. Ambiguity functions for these two cases are considered in Sections 2.4.1 and 2.4.2. In Section 2.4.3 we consider the matched processing of the separated SF signal for very large bandwidth-duration products. Simulation of range profiles in SF illumination is given in Section 2.4.4.

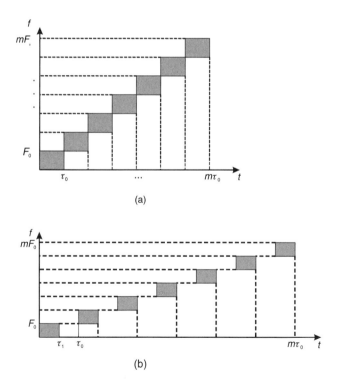

(a)

(b)

Figure 2.20 Gabor diagrams for (a) contiguous and (b) separated stepped-frequency signals.

2.4.1 Ambiguity Functions of SF Signals with Moderate Bandwidth-Duration Products

Let us use a well-known common expression of ambiguity function via the time τ and frequency F mismatches for the case of unity signal energy:

$$\dot\rho(\tau, F) = \int_{-\infty}^{\infty} \dot U\left(t + \frac{\tau}{2}\right) \cdot \dot U^*\left(t - \frac{\tau}{2}\right) \cdot e^{j2\pi Ft} dt \qquad (2.9)$$

Assuming rectangular envelopes of whole and partial pulses, we get the complex envelope of SF signals

$$\dot U(t) = \frac{1}{\sqrt{m}} \sum_{\mu=0}^{m-1} \dot U_1\left[t - \left(\mu - \frac{m-1}{2}\right)\tau_0\right] e^{j2\pi\left(\mu - \frac{m-1}{2}\right)F_0 kt} \qquad (2.10)$$

where m is the number of partial pulses, $U_1(t)$ is the envelope of partial pulse, τ_0 is the pulse repetition period, and F_0 is the frequency step. Substituting (2.10) into (2.9) we can obtain

$$\dot\rho(\tau, F) = \frac{1}{m} \sum_{\mu=0}^{m-1} \sum_{\nu=0}^{m-1} \int_{-\infty}^{\infty} \dot U_1\left[t + \frac{\tau}{2} - \left(\mu - \frac{m-1}{2}\right)\tau_0\right]$$

$$\cdot \dot U_1^*\left[t - \frac{\tau}{2} - \left(\nu - \frac{m-1}{2}\right)\tau_0\right] \qquad (2.11)$$

$$\cdot e^{j2\pi[F+(\mu-\nu)F_0]t} dt \times e^{j\pi(\mu+\nu-m+1)F_0\tau}$$

Denoting $\lambda = \mu - \nu$; $F' = F + \lambda F_0$; $\tau' = \tau - \lambda\tau_0$; $t = t' + (2\nu + \lambda - m + 1)\tau_0/2$ in (2.11), we obtain

$$\dot\rho(\tau, F) = \frac{1}{m} \sum_{\mu=0}^{m-1} \sum_{\nu=0}^{m-1} \int_{-\infty}^{\infty} \dot U_1\left(t' - \frac{\tau'}{2}\right) \dot U_1^*\left(t' - \frac{\tau'}{2}\right)$$

$$\cdot e^{j2\pi F't'} dt' \, e^{j2\pi[F'(2\nu+\lambda-m+1)\tau_0/2+(2\nu+\lambda-m+1)F_0\tau/2]} \qquad (2.12)$$

The internal integral of (2.12) is the ambiguity function of a partial rectangular pulse of duration τ_1:

$$\dot{\rho}_1(\tau', F') = \frac{\sin[\pi F'(\tau_1 - |\tau'|)]}{\pi F' \tau_1} \, \text{rect}\left[\frac{\tau'}{2\tau_1}\right] \qquad (2.13)$$

where $\text{rect}[x] = 1$ if $|x| \leq 0.5$, and $\text{rect}[x] = 0$ otherwise. Taking into account that

$$(F + \lambda F_0)(2\nu + \lambda - m + 1)\tau_0/2 + (2\nu + \lambda - m + 1)F_0\tau/2$$
$$= \left(\nu - \frac{m - 1 - \lambda}{2}\right)(F_0\tau + F\tau_0 + \lambda F_0\tau_0)$$

one obtains (2.12) in the form

$$\dot{\rho}(\tau, F) = \frac{1}{m}\sum_{\lambda}\sum_{\nu} \dot{\rho}_1(\tau - \lambda\tau_0, F + \lambda F_0)e^{j2\pi\left(\nu - \frac{m-1-\lambda}{2}\right)(F\tau_0 + F_0\tau + \lambda F_0\tau_0)}$$

$$(2.14)$$

Summation limits in (2.14) can be found from the inequalities

$$0 \leq \nu \leq m - 1, \quad \mu = \nu + \lambda, \quad 0 \leq \mu \leq m - 1, \quad |\tau - \lambda\tau_0|/2\tau_1 \leq 0.5$$
$$(2.15)$$

determined by the limited number of partial pulses and by limited partial pulse duration.

Considering these relations together, one can find the summation limits for ν and λ:

$$\frac{\lambda - |\lambda|}{2} \leq \nu \leq m - 1 - \frac{\lambda + |\lambda|}{2} \qquad (2.16)$$

$$\frac{\tau}{\tau_0} - \frac{\tau_1}{\tau_0} \leq \lambda \leq \frac{\tau}{\tau_0} + \frac{\tau_1}{\tau_0} \qquad (2.17)$$

Separated SF Signal. Assuming that $\tau_0/\tau_1 > 2$, we can see that inequalities (2.17) are satisfied when

$$\lambda = \lambda(\tau) = \frac{\tau}{|\tau|}\left\{\frac{|\tau|}{\tau_0} + \frac{1}{2}\right\} \qquad (2.18)$$

where $\{x\}$ is the whole part of x. The sum by λ in (2.14) is then reduced for each given m to a single summand. The sum by ν in (2.14) is then reduced to the geometric progression summed in the limits (2.16). The ambiguity function absolute value takes the form

$$\rho(\tau, F) = \left| \dot{\rho}_1(\tau - \lambda\tau_0, F + \lambda F_0) \frac{\sin[\pi(m - |\lambda| b]}{m \sin[\pi b]} \right| \qquad (2.19)$$

where $b = F_0\tau + F\tau_0 + \lambda F_0\tau_0$ and λ is given by (2.18).

The horizontal sections of ambiguity function (2.19) are shown using contours in Figure 2.21.

Contiguous SF Signal. Such a signal usually has moderate bandwidth-duration product. The ratio $\tau_0/\tau_1 = 1$ and inequality (2.17) are satisfied by the two whole numbers $\lambda_1 = \lambda_1(\tau) = \{\tau/\tau_0\}$ and $\lambda_2 = \lambda_2(\tau) = \lambda_1(\tau) + \tau/|\tau|$. The ambiguity function absolute value takes the form

$$\rho(\tau, F) = \left| \sum_i \dot{\rho}_1(\tau - \lambda_i\tau_0, F + \lambda_i F_0) \frac{\sin[\pi(m - |\lambda_i| b_i)]}{m \sin[\pi b_i]} \right|$$

$$(2.20)$$

where $i = 1, 2$ and $b_i = F_0\tau + F\tau_0 + \lambda_i F_0\tau_0$ [20].

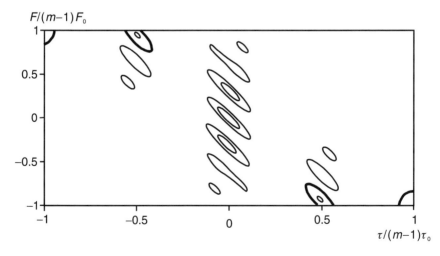

Figure 2.21 Horizontal sections of ambiguity function of separated stepped-frequency signal with moderate bandwidth-duration product.

2.4.2 Ambiguity Functions of Separated SF Signal with Very Large Bandwidth-Duration Product

The doppler effect appears in this case not only as a frequency shift but also as the frequency-time scale transformation (Section 1.1.2). The pulse repetition period τ_0 and frequency step F_0 of the received signal are shifted, becoming notably different from those τ_{00} and F_{00} of the transmitted signal. The relations between the received F_0, τ_0 and the transmitted F_{00}, τ_{00} signals' parameters are given with proportions

$$\frac{F_0}{F_{00}} = \frac{\tau_{00}}{\tau_0} = \frac{f_0}{f_{00}} = \frac{1 - v_r/c}{1 + v_r/c} \approx 1 - \frac{2v_r}{c} \qquad (2.21)$$

For a moderate bandwidth-duration product, all the frequency components of a signal possess the same doppler shifts $F = f_{00} - f_0$. For the very large bandwidth-duration product, the frequency components possess their own shifts; therefore, only the doppler frequency F for the carrier frequency will be used in calculations. Introducing the doppler shift F into carrier frequency $f_0 = f_{00} - F$, one obtains from proportions (2.21) that

$$F_0 = F_{00}(f_{00} - F)/f_{00} \quad \text{and} \quad \tau_0 = \tau_{00}f_{00}/(f_{00} - F) \qquad (2.22)$$

So, the values F_0 and τ_0 depend now on the doppler shift F for the carrier frequency. Substituting (2.22) into (2.19), we evaluate absolute value of ambiguity function in the form

$$\rho(\tau, F) = \qquad (2.23)$$

$$\left| \dot{\rho}_1\left(\tau - \lambda\tau_{00}\frac{f_{00}}{f_{00} - F}, \ F + \lambda F_{00}\frac{f_{00} - F}{f_{00}} \right) \frac{\sin\left[\pi(m - |\lambda| b)\right]}{m \sin\left[\pi b\right]} \right|$$

where

$$b = b(\tau, F) = F_{00}\tau\frac{f_{00} - F}{f_{00}} + F\tau_{00}\frac{f_{00}}{f_{00} - F} + \lambda F_{00}\tau_{00}$$

The horizontal sections of ambiguity function are shown in Figure 2.22 with parameters chosen for convenience of demonstration. The signal peak drops and dissipates for time and frequency mismatches. The horizontal sections of the ambiguity function become S-shaped due to nonequal doppler shifts of different frequency components [20].

$F/(m-1)F_0$

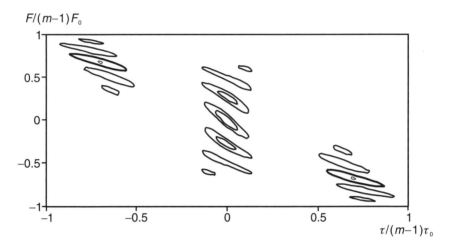

$\tau/(m-1)\tau_0$

Figure 2.22 Horizontal sections of ambiguity function of separated stepped-frequency signal with very large bandwidth-duration product.

2.4.3 Matched Processing of Separated SF Signal with Very Large Bandwidth-Duration Product

Matched processing presumes the matching of a received and an expected (reference) signal in all their parameters. Let us show some examples of matched processing realization.

Combined (Correlation and Filter) Processing for Precisely Known Radial Velocity. The simplified scheme of processing is shown in Figure 2.23. The signal is filtered and amplified by the narrowband intermediate frequency amplifier after heterodyning in the mixer. Unlike the usual matched filter processing, the constant frequency heterodyne signal is replaced with the joined SF one: (1) with steps lasting through pulse repetition periods, and (2) with central frequency differing relative to that of illumination SF signals by the intermediate frequency. The mixer, together with the phase detector controlled by the intermediate frequency reference signal, performs the correlation processing. The reference signal generator is controlled by the generator of range gates and phase shifts. It takes into account the doppler effect for the known target radial velocity. This effect appears for a short pulse as the initial phase shift and time delay of partial pulses. The gate position and heterodyne phase shift are changed in accordance with the target's motion. The phase detector has two quadrature outputs sampled by the analog-to-digital converter (ADC), stored and utilized after precursory weighting (PW) as the input signals of the DFT. Operation of this scheme is influenced by

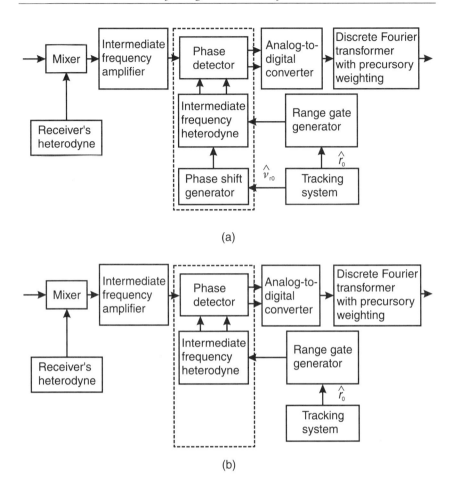

Figure 2.23 Simplified scheme of combined (correlation and filter) processing for separated stepped-frequency signal with the motion compensation implemented (a) in its analog part and (b) in its digital part.

the quality of evaluation of target movement and the signal bandwidth-duration product. The larger the signal bandwidth-duration product, the higher the required quality of movement parameter evaluation. Such evaluation can be performed both in analog and digital processing. Some examples are given below.

Filter Processing for Approximately Known Radial Velocity. Consideration of analog processing allows us to find acceptable realizations of the digital one. A restricted number of processing channels can be used to account for the unknown target movement. Each channel can introduce a different time-frequency scale. For the precise carrier frequency, taking into account the number M of these channels depends on the error Δv_r of matching of the target's actual and expected radial velocities and on the signal bandwidth-duration product $T\Delta f$. So,

$$M = kT\Delta f2\Delta v_r/c \qquad (2.24)$$

where c is the velocity of light and $k = 2$ to 5 is the ratio of compressed pulse duration to the maximum mutual time shift of its components. The scheme of a matched filter is shown in Figure 2.24. Individual pulses are taken from the output of the wideband intermediate frequency amplifier. The filter sums the outputs of multiple-tap delay lines to introduce the required stretching or compressing of the time-frequency scale.

Combined Digital Processing for Approximately Known Radial Velocity. The analog part of digital correlation processing [Figure 2.23(a)], denoted by the dashed line, is simplified using an intermediate frequency heterodyne without precise phase control [Figure 2.23(b)]. Phase shifts caused by an expected radial velocity (Figure 2.23) are introduced by subtraction of phase $(4\pi\Delta v_{rk}f_i t_i/c)$ from each i th pair of digital quadrature samples. The required number of processing channels $k = 1, 2, \ldots, M$; (2.24) will determine steps in radial velocity v_r alignment.

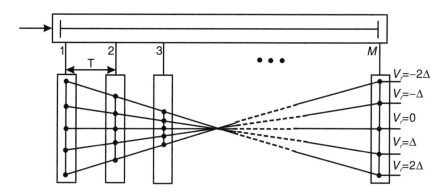

Figure 2.24 Scheme of the multichannel filter processing of separated stepped-frequency signal for approximately known radial velocity as a basis to implement the digital processing.

2.4.4   Simulated and Experimental RPs for Separated SF Illumination

The quality of an RP is affected by many factors, such as rotational modulation and yaws of a target, which have to be considered. Combined correlation and filter processing is presumed to provide matching of reference pulse phases, time delays, pulse repetition periods, and frequency steps. The filtering is reduced to DFT of signal samples and is performed after the correlation processing.

Simulated RPs neglecting a target's rotational modulation and yaws were considered for different mismatches of actual and expected target radial velocity both for point and extended targets.

Figure 2.25 shows the processed signals (RPs) for the point target and different radial velocity mismatches: 0 m/s [Figure 2.25(a, c)], 45 m/s [Figure 2.25(b, d)]. The SF signal has the f_0 = 3 GHz carrier, consists of N = 64 pulses of τ_0 = 0.5 μs duration with pulse repetition frequencies of 20 kHz [Figure 2.25(a, b)] and of 5 kHz [Figure 2.25(c, d)]. The frequency steps are F_0 = $1/\tau_0$ = 2 MHz. The interval of range unambiguous measurements r_{un} = $c/2F_0$ = 75m is shown. For no mismatch of radial velocity, the processed signal exhibits its potential resolution. Radial velocity mismatch leads to a decrease in the processed pulse resolution and the time shift. The 45-m/s mismatch corresponds to coefficient k = 2 in (2.24).

Figures 2.26 and 2.27 show the RPs of Tu-16-type (Figure 2.26) and An-26-type (Figure 2.27) aircraft for zero radial velocity mismatch without accounting for disturbing factors (turbine modulation and the target yaws). Decrease of PRF from 20 kHz [Figures 2.26(a) and 2.27(a)] to 5 kHz

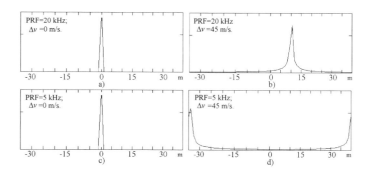

Figure 2.25   Simulated range profiles obtained with separated SF illumination for a point target with two PRFs of (a, b) 20 kHz and (c, d) 5 kHz with various radial velocity mismatches of (a, c) zero and (b, d) 45 m/s. Wavelength is 3 cm, number of pulses is 64, pulse duration is 0.5 μs, frequency step is 2 MHz, and bandwidth is 128 MHz.

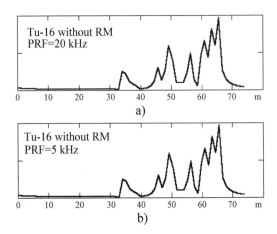

Figure 2.26 Simulated range profiles obtained with separated SF illumination for Tu-16-type aircraft in the absence of rotational modulation for two PRFs of (a) 20 kHz and (b) 5 kHz with zero radial velocity mismatch. Wavelength is 3 cm, number of pulses is 64, pulse duration is 0.5 μs, frequency step is 2 MHz, bandwidth is 128 MHz, and target aspect is 20° from nose.

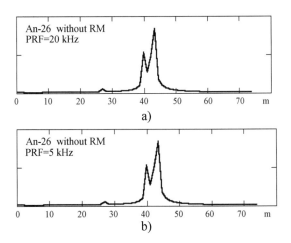

Figure 2.27 Simulated range profiles obtained with separated SF illumination for An-26-type aircraft in the absence of rotational modulation for two PRFs of (a) 20 kHz and (b) 5 kHz with zero radial velocity mismatch. Wavelength is 3 cm, number of pulses is 64, pulse duration is 0.5 μs, bandwidth is 128 MHz, and target aspect is 20° from nose.

[Figures 2.26(b) and 2.27(b)] does not significantly change the RPs in the absence of a disturbing factor.

Simulated RP Accounting for Target's Rotational Modulation and Neglecting Yaws. Rotating systems of a target (turbine compressors and propellers) affect the processing of SF signals with very large bandwidth-duration product, especially for low PRF and in the presence of propeller modulation. The spectrum of a backscattered signal is more dense for propeller modulation than for turbine modulation, as can be seen from Figure 3.5. Range profiles for this case are shown in Figure 2.28 for the Tu-16-type turbojet and in Figure 2.29 for the An-26 turbo-prop aircraft. Figures 2.28(a) and 2.29(a) correspond to 20 kHz, and Figures 2.28(b) and 2.29(b) correspond to 5-kHz PRF.

Comparing the results for separated SF signals with those for chirp we can see that for the wavelength of 3 cm the RPs are distorted by rotational modulation. The distortions are especially significant for the 5-kHz PRF and are smaller for the 20-kHz PRF.

The level of rotational modulation depends significantly on the aspect angle of a target. It is interesting, therefore, to simulate an outward flight of a target (see Section 3.3), where the level of rotational modulation is usually lower than that for an inward flight. The simulated RP of

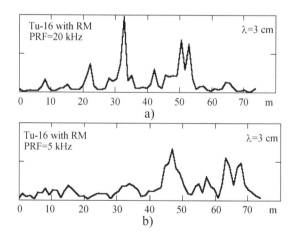

Figure 2.28 Simulated range profiles obtained with separated SF illumination for Tu-16-type aircraft in the presence of rotational modulation for two PRFs of (a) 20 kHz and (b) 5 kHz with zero radial velocity mismatch. Wavelength is 3 cm, number of pulses is 64, pulse duration is 0.5 μs, bandwidth is 128 MHz, and target aspect angle is 20° from nose.

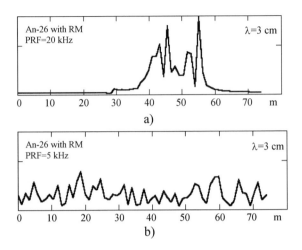

Figure 2.29 Simulated range profiles obtained with separated SF illumination for An-26-type turbo-prop aircraft in the presence of rotational modulation for pulse repetition frequencies of (a) 20 kHz and (b) 5 kHz with zero radial velocity mismatch. Wavelength is 3 cm, number of pulses is 64, pulse duration is 0.5 μs, bandwidth is 128 MHz, and target aspect angle is 20° from nose.

Tu-16-type aircraft flying away from the radar with the aspect of 135° from the nose is shown in Figure 2.30 in a noise background. The parameters of the separated SF illumination signal are as follows: wavelength is 3 cm, number of pulses is 128, frequency step is 2 MHz, bandwidth is 256 MHz, PRF is 20 kHz.

Experimental RPs.   We can compare the latter simulation results with experimental ones of [21]. In [21] the 256-MHz stepped-frequency illumination signal at 3-cm wavelength consisting of 128 pulses with PRF of 20 kHz and frequency step of 2 MHz was used. By means of this signal, the RP of a B-727 turbojet aircraft was observed with aspect angle of 135° from the nose (Figure 2.31). As follows from simulation, the rotational modulation is not very significant for this aspect angle.

We can see that the length of the RPs in Figures 2.30 and 2.31 are close to each other in similarity of their actual dimensions. The shape of the RPs can be different due to different structures of the two aircraft. However, it is possible that for high-range resolution the limitation of the chosen simulation method emerges (see also Section 2.5.4 and Chapter 7). We have no sufficient data to clarify this subject yet.

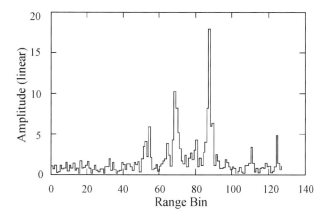

Figure 2.30 Simulated range profile obtained with separated stepped-frequency illumination for Tu-16-type turbojet aircraft with turbine modulation; the PRF is 20 kHz, wavelength is 3 cm, number of pulses is 128, pulse duration is 0.5 μs, bandwidth is 256 MHz, and target aspect angle is 135° from nose.

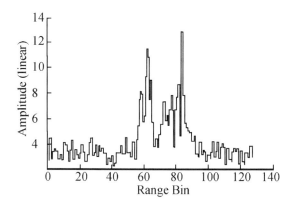

Figure 2.31 Experimental range profile obtained with separated SF illumination for B-727 turbojet aircraft with turbine modulation. The PRF is 20 kHz, wavelength is 3 cm, number of pulses is 128, pulse duration is 0.5 μs, bandwidth is 256 MHz, and target aspect angle is 135° from nose. (*Source:* [21, Figure 5] © 1996 IEEE, reprinted with permission.)

Comparison of Separated SF Illumination with Chirp.   Separated SF illumination attracts attention due to relative receiver simplicity, but acquiring quality RPs with separated SF illumination poses more difficulties than for chirp. The difficulties are caused by the more time-extended ambiguity function.

The influence of such a distorting factor as rotational modulation is therefore increased, especially for not very high PRF and high carrier frequencies.

2.5 Target's 2D Images

The technology of 2D moving target imaging or inverse synthetic aperture radar (ISAR) technology attracts great attention [1, 2, 9, 22–24]. Let us consider first the backscattered signal model (Section 2.5.1) and the possibilities of ISAR processing on the basis of one or several reference elements (Section 2.5.2). We consider then ISAR processing on the basis of the Wigner-Ville (WV) transform (Section 2.5.3). Some of our image simulation examples are given in Section 2.5.4. Some difficulties are considered as the discussion progresses.

2.5.1 Models of Backscattered Signal and Processing Variants for ISAR

Models of the backscattered signal are considered here for the case of target motion in the "radar-target" plane. The reflecting elements are not supposed to escape from their range cells. Expected signal complex amplitudes for these cells may be written as a summation of complex amplitudes of signals backscattered by a set of target elements $k = 1, 2, \ldots, M$ entered into the same range resolution cell and unresolved in cross range [22, 23]:

$$X(t) = \sum_{k=1}^{M} A_k \exp\left\{ -j\left(\frac{4\pi}{\lambda} [r(t) + \xi_k \cos\theta(t) + \eta_k \sin\theta(t)] + \psi_k \right) \right\}$$

$$(2.25)$$

Here, A_k is the amplitude of the signal scattered by the kth target element; ξ_k, η_k are its longitudinal and transverse coordinates relative to the chosen origin; $r(t) = r_0 + v_r t + \frac{1}{2} a_r t^2 + \ldots$ is the law of longitudinal short-term target motion; $\theta(t) = \Omega t + \frac{1}{2} a_\Omega t^2 + \ldots$ is the law of angular short-term target motion relative to the line-of-sight; ψ_k is the initial phase. Additional parameters are introduced here: the initial radial target velocity v_r; the angular velocity Ω defined relative to the line-of-sight; the radial a_r and angular a_Ω accelerations. Using approximations $\cos\theta(t) \approx 1 - \frac{1}{2}\theta^2(t)$ and $\sin\theta(t) \approx \theta(t)$ for the short-term target motion $\theta(t) \ll 1$, we obtain

$$X(t) \approx \sum_{k=1}^{M} A_k \exp\left\{-j\left(\frac{4\pi}{\lambda}\left[r_0 + (v_r + \eta_k\Omega)t\right.\right.\right. \tag{2.26}$$

$$\left.\left.\left. + \frac{1}{2}(a_r - \xi_k\Omega^2 + \eta_k a_\Omega)t^2\right] + \psi_k\right)\right\}$$

Another form of (2.26) is

$$X(t) = \sum_{k=1}^{M} A_k \exp\left\{-j\left(2\pi\left[\int_0^t F_k(s)ds + 2r_0/c\right] + \psi_k\right)\right\} \tag{2.27}$$

In (2.27) the doppler frequency of the kth target element varying in time was introduced

$$F_k = F_k(t) = \frac{2}{\lambda}\left[v_r + \eta_k\Omega + (a_r - \xi_k\Omega^2 + \eta_k a_\Omega)t\right] \tag{2.28}$$

with the constant derivative

$$F_k' = F_k'(t) = \frac{2}{\lambda}(a_r - \xi_k\Omega^2 + \eta_k a_\Omega) \tag{2.29}$$

The simplest variant of model (2.28) and (2.29) corresponds to $a_\Omega = 0$, $a_r = 0$, $2|\xi_k|\Omega^2/\lambda \ll 1$.

Possible Variants of ISAR Processing. As we can see, the number of unknown parameters of (2.26) through (2.29) is too great for simple evaluation and for successive realization of matched processing. Several kinds of autocorrelation processing and various problem simplifications are therefore considered. The autocorrelation processing is partly realized using reference signals back-scattered from dominating target elements [1, 8, 9, 22], or by the WV transform variants [24].

2.5.2 ISAR Processing on the Basis of Reference Target Elements

ISAR processing using a single reference target element [9] is accepted for a nonmaneuvering target ($a_\Omega = 0$, $a_r = 0$), uniformly moving in a direction that differs from the radial. In the process of motion the target rotates relative to the line-of-sight with some angular velocity $\Omega \neq 0$, evaluated approximately from trajectory processing. For $2|\xi_k|\Omega^2/\lambda \ll 1$, the value of $F_k' \approx 0$. Then, a series of the changing target range profiles can be obtained in the time interval T by sequential wideband illumination. An intensive and steady dominating element of profile can be selected algorithmically. It can be supposed for wide bandwidths that the dominating element corresponds to a single resolved backscatterer. The sample of the dominant element can be used as the reference. Its phase is subtracted from the phases of all the other samples of the profile. The doppler frequencies F of these elements are then transformed into differential doppler frequencies $F_{dif} = F - F_{ref}$, where $F_{ref} = F_1$ is the doppler frequency of the reference element $k = 1$. The differential doppler frequencies are equivalent to the nondifferential ones but conditioned by the target rotation around the axis crossing the reference target element and normal to the flight plane. If the origin of the coordinate system is associated with this element, the differential doppler frequencies of all the other elements with coordinates ξ_k, η_k are

$$F_{k_{dif}} = 2\Omega\eta_k/\lambda \tag{2.30}$$

An analogous result can be obtained from (2.28). Estimation of unknown coordinate η_k and amplitude A_k consists, therefore, of the Fourier transform of identical elements ξ_k = const of range profile series and of measuring the frequency and amplitude of each spectrum component. Frequency resolution $1/T$ in matched processing determines the measure $\Delta\eta$ of cross-range resolution

$$\Delta\eta = \lambda/2\Omega T = \lambda/2\Delta\alpha \tag{2.31}$$

where $\Delta\alpha = \Omega T$ is the angle of target rotation relative to the line-of-sight for the observation time T [1, 9]. To obtain the cross-range resolution $\Delta\eta = 1$ to 3m for the wavelength $\lambda = 3$ to 10 cm and the coherent integration time of 0.5 to 1 sec (the time of 1 sec corresponds to particularly perfect weather conditions), the value of $\Delta\alpha$ must be 0.05 to 0.005 radian or 3 to 0.3° and the value of $|\Omega|$ must be 0.1 to 0.005 rad/s. With straightforward uniform motion, the absolute value $\Omega = |v\sin\alpha|/r$, where v is the target velocity. So, for the $\alpha = \pi/2$ to $\pi/4$, $v = 0.2$ to 0.5 km/s and $\Omega = 0.1$ to 0.005 rad/s the possible range of 2D imaging is $r < 1.4$ to 100 km. In many, but not all, important practical cases this range is too small.

ISAR processing using three reference target elements has to account for the elements of target maneuver ($a_{\Omega\Sigma} \neq 0$, $a_r \neq 0$) and arbitrary values

$2|\xi_k|\Omega^2/\lambda$ of the model (2.27), which were neglected above. After the previous kind of processing [22]:

- The longitudinal coordinates ξ_k of all the target elements are known. The dominant element $k = 1$ became the origin $\xi_1 = \eta_1 = 0$ of $O_{tg}\xi\eta\zeta$ coordinate system.
- Two additional reference elements can be chosen. With maneuvering, their actual coordinates η_2, η_3 depending of angular velocity Ω are unknown, but their ratio η_2/η_3 is independent of Ω and can be evaluated.

New information about the derivatives of differential doppler frequencies $(F_{k\,\mathrm{dif}})' = F'_k - F'_1$, $k = 2$, 3 can be obtained by tracking the target elements. A system of equations for model (2.26) through (2.29) can be obtained from (2.29):

$$-\xi_2\Omega^2 + \eta_2 a_\Omega = \frac{\lambda}{2}(F'_2 - F'_1)$$

$$-\xi_3\Omega^2 + (\eta_3/\eta_2)\eta_2 a_\Omega = \frac{\lambda}{2}(F'_3 - F'_1)$$

It defines the Ω and $\eta_2 a_\Omega$ values, which can be used to compensate for the maneuver components of the model. The value of Ω determines in principle the cross-range scale. But with limited observation time and straightforward uniform motion of the target, it is difficult to estimate the doppler frequency derivatives in most cases.

2.5.3 ISAR Processing on the Basis of the WV Transform

WV Transform [24]. Two heuristic approaches lead to this transform. One is connected with the Wiener-Khinchin equation for stationary random processes; the other is connected with the Woodward time-frequency ambiguity function of radar signals.

Let us begin with the first approach, connected with the stationary random processes $Y(t)$ having time-independent correlation function $R(\tau) = E[Y(t + \tau/2)Y^*(t - \tau/2)]$, where $E[x]$ is used as the expectation symbol of the random value x. The power spectrum in this case is exactly defined by the Wiener-Khinchin equation

$$G(f) = \int\limits_{-\infty}^{\infty} R(\tau) \exp(-j2\pi f\tau)d\tau \qquad\qquad (2.32)$$

To evaluate approximately the power spectrum $G(f, t)$ of a quasi-stationary process with correlation function $R(\tau, t) = E[Y(t + \tau/2)Y^*(t - \tau/2)]$, moderately depending on time, one can use analogous equation

$$G(f, t) = \int\limits_{-\infty}^{\infty} R(\tau, t) \exp(-j2\pi f\tau)d\tau \qquad\qquad (2.33)$$

As usual, there are no great numbers of realizations $Y(t)$ to provide the operation $E[\cdot]$ of their averaging. Therefore, time averaging must also be used for $R(\tau, t)$ and $G(f, t)$ moderately depending on time. But for a stronger dependence on time, independent time averaging becomes impossible. That leads to rejection of time averaging and to the use of the WV transform, with time averaging being realized in the process of spectral transform

$$G(f, t) = \int\limits_{-\infty}^{\infty} Y(t + \tau/2)Y^*(t - \tau/2) \exp(-j2\pi f\tau)d\tau \qquad (2.34)$$

The second heuristic approach to the WV transform uses the concept and properties of the Woodward time-frequency ambiguity function of radar signals without noise. In expression (2.5) for the ambiguity function, one can formally replace the signal $X(t)$ without noise with the signal $Y(t)$, containing the noise, that leads to (2.34). Although the result of such substitution has an analogy with optimal processing, it is not optimal.

Advantage, Deficiency, and Development of the WV Transform. The advantage of the WV transform is some processing simplification with target maneuver. The great deficiency is the spectrum distortions because of the nonlinear processing. So, when the signal contains more than one frequency component, the WV transform creates spurious cross-term interference, occurring at random positions of the time-frequency plane. A kind of cross-term suppression based on the 2D Gabor expansion of the WV transform into the time-frequency distribution series (TFDS) was therefore developed [24]. The TFDS simulation is beyond the scope of this book.

2.5.4 Examples of 2D Image Simulation

Let us consider image simulation examples for various target orientations in four cases: (1) with target rotational uniform motion; (2) with linear uniform motion without yaws and practically without rotational modulation; (3) with linear uniform motion without yaws and with significant rotational modulation; and (4) with translational motion with yaws and practically without rotational modulation. The image contrast was artificially improved.

Example of Simulation for Uniform Rotational Target Motion. A simulated 2D image of a B-52-type aircraft and the direction of its illumination are shown in Figure 2.32. The target is assumed to have uniform rotation in aspect (course-aspect angle) from 95.86° to 95° around the target coordinate system origin without translational motion. The target roll- and pitch-aspect angles are supposed to be constant. We chose them artificially to have almost all the bright elements unshadowed by others. The illumination is performed on a carrier wavelength of 3 cm with Gaussian chirped pulses of 160-MHz deviation, each of them providing resolution in range of about 1m. The number of pulses processed to obtain the image is 128. The target rotation

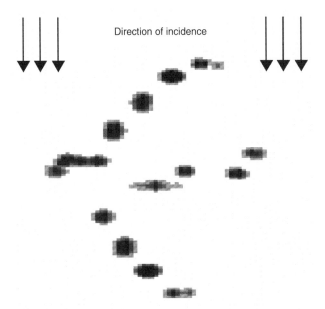

Figure 2.32 Simulated 2D image of B-52-type aircraft rotating in aspect from 95.86° to 95°. The carrier wavelength λ is 3 cm; the range and cross-range resolutions are equal to 1m.

angle was chosen by (2.31) to provide cross-range resolution of 1m. The most principal bright points are not shadowed in this case, so the target's 2D image could be easily interpreted.

Examples of Simulation for Linear Uniform Target Motion Without Yaws and Practically Without Rotational Modulation. The simulated 2D images of a B-52-type aircraft flying straightforwardly with constant velocity and the directions of its illumination are shown in Figure 2.33(a) and (b) for two carrier wavelengths. The target flies without random disturbances, and rotational modulation is negligible. The initial aircraft distance is 15 km, its

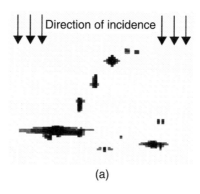

(a)

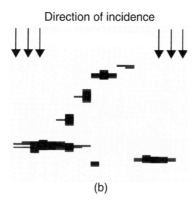

(b)

Figure 2.33 Simulated 2D images of B-52-type aircraft for its straightforward flight with the aspect 95° without yaws and practically without turbine modulation for two cases: (a) carrier wavelength λ is 3 cm, cross-range resolution is 1m; and (b) carrier wavelength λ is 10 cm, cross-range resolution is 3m. The target initial range is 15 km, the target altitude is 5000m, the target velocity is 800 km/h, and range resolution is 1m.

flight altitude is 5000m and velocity is 800 km/h. A motion compensation procedure with only one reference point is sufficient. The course is 95°. Bright elements of the left wing are shadowed by the target body. The illumination signal consists, as in the previous case, of 128 Gaussian chirped pulses with frequency deviation of 160 MHz and the PRF of 125 Hz. Under these conditions the observation time of 1 sec provides target rotation by 0.9° and cross-range resolution of about 1m for the carrier wavelength λ = 3 cm [Figure 2.33(a)]. For the longer carrier wavelength λ = 10 cm, the cross-range resolution is reduced to about 3m [Figure 2.33(b)].

Examples of Simulation for Linear Target Motion with Rotational Modulation. Rotational (turbine) modulation may distort the 2D image, as illustrated in Figure 2.34(a). The initial target aspect is chosen here equal to 115° to increase the turbine effect. The carrier wavelength λ = 3 cm and the remaining parameters are the same as in the previous case. Together with some decrease of cross-range resolution, the additional points occur on the image complicating its interpretation. The influence of rotational modulation decreases on the carrier wavelengths $\lambda \geq 10$ cm [Figure 2.34(b)] and in the cases of a higher PRF [Figure 2.34(c)].

Example of Simulation for Translational Target Motion with Yaws and Practically Without Rotational Modulation. The target's random motion limits the intervals of the coherent integration and cross-range resolution, especially at large distances. Influence of target yaws on the 2D image is illustrated in Figure 2.35 for cloudy weather (see Section 1.4). The carrier wavelength and other parameters are the same as for Figure 2.33(a).

Some Conclusions from the Results of Analysis and Simulation. A high-quality 2D image contains a body of information, but it is difficult to obtain such an image for the usual ranges of radar operation. Considering the case of the chirp signal, we illustrated above some difficulties of obtaining high-quality images: the shadowing of the image parts and the possible image distortions due to the rotational modulation and yaws of target. Taking into account the conclusions of Section 2.4, one can note that in the case of the separated SF signal, the difficulties may become even greater.

One can also see that the simulation method used above may be too rough in the case of very high range and cross-range resolution, because it yields as yet insufficient numbers of the image elements. This does not influence, however, the essence of the conclusions formulated above.

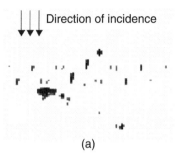

(a)

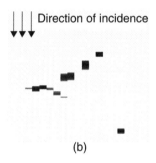

(b)

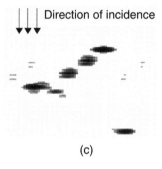

(c)

Figure 2.34 Simulated 2D images of B-52-type aircraft for straightforward flight with an aspect of 115° without yaws, but with the turbine modulation being more explicit than in Figure 2.32, for three cases: (a) carrier wavelength λ is 3 cm and PRF is 125 Hz; (b) carrier wavelength λ is 10 cm and PRF is 125 Hz; and (c) carrier wavelength λ is 3 cm and PRF is 1000 Hz. The remaining parameters are the same as for Figure 2.32(a).

Figure 2.35 Simulated 2D image of B-52-type aircraft for translational motion with yaws and practically without turbine modulation. The carrier wavelength λ is 3 cm and the remaining parameters are the same as for Figure 2.32(a).

References

[1] Wehner, D. R. *High Resolution Radar*, Second Edition, Norwood, MA: Artech House, 1994.

[2] Rihaczek, A. W., and S. I Hershkowitz, *Radar Resolution and Complex-Image Analysis*, Norwood, MA: Artech House, 1996.

[3] Shirman, Y. D., *Resolution and Compression of Signals*, Moscow: Sovetskoe Radio Publishing House, 1974 (in Russian).

[4] Shirman, Y. D., et al. "On the First Super-Wideband Radar Investigations in Soviet Union," *Radiotehnika i Electronika*, Vol. 36, January 1991, pp. 96–100 (in Russian).

[5] Shirman, Y. D., et al. "Aerial Target Backscattering Simulation and Study of Radar Recognition, Detection and Tracking," *IEEE Int. Conf. Radar-2000*, May 2000, Washington, DC, pp. 521–526.

[6] Shirman, Y. D., et al. "Study of Aerial Target Radar Recognition by Method of Backscattering Computer Simulation," *Proc. Antenna Applications Symp.*, September 1999, Allerton Park Monticello, Illinois, pp. 431–447.

[7] Chamberlain, N., E. Walton, and E. Garber, "Radar Target Identification of Aircraft Using Polarization – Diverse Features," *IEEE Trans.*, AES-27, January 1991, pp. 58–66.

[8] Shirman, Y. D., et al. "Methods of Radar Recognition and Their Simulation," *Zarubeghnaya Radioelectronika—Uspekhi Sovremennoi Radioelectroniki*, No.11, November 1996, Moscow, pp. 3–63

[9] Steinberg, B. D., "Microwave Imaging of Aircraft," *Proc. IEEE*, Vol. 76, December 1988, pp.1578–1592.

[10] Woodward, P. M. *Probability and Information Theory with Applications to Radar*, Oxford: Pergamon, 1953; Norwood, MA: Artech House, 1980.

[11] Rihaczek, A. W., *Principles of High-Resolution Radar*, New York: McGraw-Hill, 1969; Norwood, MA: Artech House, 1996.

[12] Shirman Y. D., and V. N. Golikov, *Foundations of Theory of Radar Signals' Detection and Their Parameter Measurement*, Moscow: Sovetskoe Radio Publishing House, 1963 (in Russian).

[13]   Shirman, Y. D., S. P. Leshenko, and V. M. Orlenko, "About the Simulation of Aerial Target Backscattering and Its Use in Radar Recognition Engineering," *Vestnik Moskovskogo Gosudarstvennogo Tehnicheskogo Universiteta (Radioelektronika)*, No. 4, 1998, pp. 14–25 (in Russian).

[14]   Shirman, Y. D., "Method of Radar Resolution Enhancement and the Device for its Realization," Author's Certificate No.146803 on July 25, 1956 Application, *Bulletin of Inventions*, No. 9, 1962 (in Russian).

[15]   Bromley, R. A., and B.E. Callan, "Use of Waveguide Dispersive Line in an FM Pulse Compression System," *Proc. IEE*, Vol. 114, September 1967, pp. 1213–1218.

[16]   Caputi, W. J., "Stretch: A Time-Transformation Technique," *IEEE Trans.*, AES-7, No. 2, March 1971, pp. 269–278.

[17]   Hudson, S., and D. Psaltis, "Correlation Filters for Aircraft Identification from Radar Range Profiles," *IEEE Trans.*, AES-29, July 1993, pp. 741–748.

[18]   Schiller, J., "Non-Cooperative Air Target Identification Using Radar," *RTO Meeting Proceedings*, Vol. 6, April 1998.

[19]   Pingping, L., L. Guochan, and H. Huai, "A S-Band Inverse Synthetic Aperture Radar System," *Proc. Chinese Intern. Radar Conf.*, CIRC-96, Beijing, October 1996, pp. 251–253.

[20]   Orlenko, V. M., and Y. D. Shirman, "Ambiguity Bodies of Frequency-Manipulated Signals," *Collection of Papers*, Issue 3, 2000, Moscow, Radiotekhnika Publishing House, pp. 5–64 (in Russian).

[21]   Zyweck, A., and R. E. Bogner, "Radar Target Classification of Commercial Aircraft," *IEEE Trans.*, AES-32, April 1996, pp. 598–696.

[22]   Wang, Y., Y. Ling, and V. Chen, "ISAR Motion Compensation Via Adaptive Joint Time-Frequency Technique," *IEEE Trans.*, AES-34, April 1998, pp. 670–677.

[23]   Shirman, Y. D., (ed.). "Inverse and Combined Aperture Synthesis." In *Handbook: Electronic Systems: Construction Foundations and Theory*, Second edition, Section 18.10, Kharkov (printing in Russian).

[24]   Chen, V., and S. Qian, "Joint Time-Frequency Transform for Radar Range-Doppler Imaging," *IEEE Trans.*, AES-34, April 1998, pp. 486–499.

3

Review and Simulation of Recognition Features (Signatures) for Narrowband Illumination

Narrowband illumination still remains the most widely used kind of radar illumination. Therefore, the signal signatures of narrowband illumination must be considered carefully [1–8], although these signatures provide less recognition information than those of wideband illumination. The possible signatures are enumerated in Section 3.1 and are considered in Sections 3.2 through 3.4 in more detail.

3.1 Signal Signatures Used in Narrowband Illumination

Signal signatures used in narrowband illumination are:

- RCS and other parameters of the PSM [1, 5–10];
- Rotational modulation spectra [1–4, 7, 8, 11–15];
- Correlation factors of fluctuation via frequency diversity of two or more reflected signals at close carrier frequencies [6–8].

See also Section 4.1.3 about the trajectory signatures (altitude, velocity, etc.).

111

3.2 RCS and Other Parameters of PSM

Let us consider the above-mentioned signatures, the results of their simulation, and the comparison of these results with available experimental data.

3.2.1 RCS in Narrowband Illumination and Its Simulation

RCS in narrowband illumination contains limited, but sometimes useful, recognition information about a target. The RCS value in decibels (relative to $1\ m^2$) can be estimated experimentally by the equation

$$\sigma_{\text{tg}} = 10\lg Q + 40\lg R - 10\lg W + \Delta, \text{dB} \tag{3.1}$$

Here, Q is the signal-to-noise energy ratio, R is the target range (in meters), and W is the radar potential (in meters squared)

$$W = E_{\text{rad}}G_{\text{t}}(\epsilon, \beta)A_{\text{r}}(\epsilon, \beta)/(4\pi)^2 N_0, \text{m}^2 \tag{3.2}$$

where, in its turn, E_{rad} is the radiated energy (taking into account the estimated losses in a piece of hardware) in J = W/Hz, ϵ and β are the target angle coordinates, $G_{\text{t}}(\cdot)$ is the gain of radar transmitting antenna, $A_{\text{r}}(\cdot)$ is the effective aperture area of the radar receiving antenna, and N_0 is the spectral density of noise in W/Hz = J. The value of Δ, dB in (3.1) corresponds approximately to additional losses in the propagation media and in the transmitting and receiving systems omitted in the calculations.

Several successive illuminations are necessary to smooth the RCS fluctuations (see the shaded zone between the dashed lines of Figure 2.9 showing the normalized standard deviations of RCS estimates for various targets). Unaccounted-for instabilities of transmitter power and receiver sensitivity lead to the measurement errors even in conditions of power and sensitivity monitoring. They must be accounted for as much as possible. Unaccounted-for instabilities of propagation conditions, especially for the rainy weather and low-altitude targets, are frequently inevitable. It is usually expedient, however, to use available narrowband RCS information with a definite weight because of the simplicity of obtaining it in the surveillance mode. The information about the target obtained from wideband illumination can be supplemented with that obtained from narrowband illumination. Such information can help to discriminate a small-sized passive decoy (with the Luneburg lens, for instance) from the small-sized missile, as well as to discriminate, presumably, a large-sized stealth aircraft from a large-sized nonstealth one.

Simulated and Experimental RCS. Examples of simulated statistical distributions of RCS for the targets of various types are shown in Figure 3.1. Such statistical distributions for vertical, horizontal, and cross-polarizations for the Tu-16-type aircraft are shown in Figure 3.2. It is seen that evaluation of averaged RCS gives us some information about the target dimensions (large, medium, small).

In Figure 3.3 we show the simulated mean RCS of the F-15-type aircraft via the 10° azimuth-aspect sectors in L, S, and X radar bands. All the RCS values having the probability density above the 0.8 level are plotted in dB relative to the maximum (broadside) mean value. The line segments are therefore limited by the 0.8 level of corresponding probability density. The mean RCSs for each sector and each band are shown by the circlets.

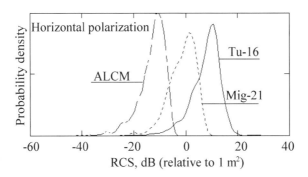

Figure 3.1 Simulated statistical distributions of RCS for the Tu-16- (solid line) and Mig-21-type (dotted line) aircraft and ALCM-type missile (dashed line) in L-band for horizontal polarization.

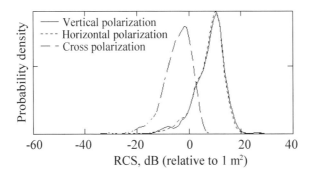

Figure 3.2 Simulated statistical distributions of RCS for the vertical (solid line), horizontal (dotted line), and cross (dashed line) polarizations in L-band for the Tu-16-type aircraft.

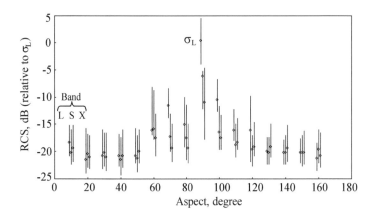

Figure 3.3 Simulated mean RCS for the F-15-type jet fighter in three radar bands.

Figure 3.3 is given by the analogy with Figure 3.3.2 of [9] that represent the corresponding experimental dependency for a jet fighter. The coincidence of Figure 3.3 and Figure 3.3.2 of [9] data is not complete, and that can be ascribed to simulation defects and to the difference in aircraft types. But the structures of dependencies are close to each other. The better coincidence of simulated and experimental data was observed for the Mig-21-type aircraft when disagreement of simulated and experimental RCS did not exceed 2.5 dB for the majority of aspect angles. Models for statistical distributions of RCSs essential for the theory of targets detection will be considered in Chapter 6.

3.2.2 Other Parameters of the Polarization Scattering Matrix and Their Simulation

Functions of the PSM Elements as the Recognition Signatures. Various signatures can be built by the heuristic combination [5, 6] of the PSM elements of (1.3) with and without representation of the PSM through the diagonal matrix (1.6).

Let us begin with the simplest signatures of a conducting sphere and a conducting dipole. The wavelength is assumed to be small in relation to the sphere radius and large in relation to the dipole thickness. The diagonal matrix of PSM's square roots from eigenvalues in a linear basis has two equal nonzero elements for the sphere and only one for the dipole.

Let us go on to the signatures that are similar to the considered range-polarization profiles (Section 2.3.1). For the narrowband illumination we have been limited to only one sample $m = 1$ of the coherent RPP (2.8). Such a polarization profile appears for a given polarization of the illumination signal of monostatic radar as $\dot{\mathbf{X}} = \left\| X' \ X'' \exp(j\beta) \right\|^{\mathrm{T}}$. Its modulus is $\left| \dot{\mathbf{X}} \right| = \sqrt{(X')^2 + (X'')^2}$ and its normalized form is $\dot{\mathbf{X}}_n = \dot{\mathbf{X}} / \left| \dot{\mathbf{X}} \right|$. The correlation procedure $Z = \left| \dot{\mathbf{Y}}^{*\mathrm{T}} \dot{\mathbf{X}} \right|$ analogous to the procedure of Section 2.3.1 or the normalized one can then be used.

Let us show that the processing procedure described in Section 2.3.1 is connected with the evaluation of the PSM elements. If the illumination signal has horizontal polarization and reception is provided at horizontal and vertical linear polarizations, this vector of the expected signal $\dot{\mathbf{X}}$ is equal to

$$
\dot{\mathbf{X}}_{\mathrm{h}} = \left\| \begin{matrix} \sqrt{\sigma_{11}} \cdot e^{j\varphi_{11}} & \sqrt{\sigma_{12}} \cdot e^{j\varphi_{12}} \\ \sqrt{\sigma_{21}} \cdot e^{j\varphi_{21}} & \sqrt{\sigma_{22}} \cdot e^{j\varphi_{22}} \end{matrix} \right\| \left\| \begin{matrix} K \\ 0 \end{matrix} \right\| = \left\| X'_{\mathrm{h}} \ X''_{\mathrm{h}} \exp(j\beta_{\mathrm{h}}) \right\|^{\mathrm{T}}
$$

(3.3)

Parameter K can be chosen from the normalization condition for vector $\dot{\mathbf{X}}_{\mathrm{h}}$ or vector $\dot{\mathbf{X}}$ defined below by (3.5).

In the case when the illumination signal has vertical polarization and the reception is provided at horizontal and vertical linear polarizations, the vector of the expected signals is equal to

$$
\dot{\mathbf{X}}_{\mathrm{v}} = \left\| \begin{matrix} \sqrt{\sigma_{11}} \cdot e^{j\varphi_{11}} & \sqrt{\sigma_{12}} \cdot e^{j\varphi_{12}} \\ \sqrt{\sigma_{21}} \cdot e^{j\varphi_{21}} & \sqrt{\sigma_{22}} \cdot e^{j\varphi_{22}} \end{matrix} \right\| \left\| \begin{matrix} 0 \\ K \end{matrix} \right\| = \left\| X'_{\mathrm{v}} \ X''_{\mathrm{v}} \exp(j\beta_{\mathrm{v}}) \right\|^{\mathrm{T}}
$$

(3.4)

If the conditions of coherent integration of information are fulfilled, we can introduce the associated signature

$$
\dot{\mathbf{X}} = \left\| \dot{\mathbf{X}}_{\mathrm{h}} \ \dot{\mathbf{X}}_{\mathrm{v}} \right\|^{\mathrm{T}} = \left\| X'_{\mathrm{h}} \ X''_{\mathrm{h}} \exp(j\beta_{\mathrm{h}}) \ X'_{\mathrm{v}} \ X''_{\mathrm{v}} \exp(j\beta_{\mathrm{v}}) \right\|^{\mathrm{T}} \quad (3.5)
$$

As it was in the case of RPs (Section 2.2), a cluster of polarization profiles $\dot{\mathbf{X}}$ (or $\dot{\mathbf{X}}_{\mathrm{h}}$, or $\dot{\mathbf{X}}_{\mathrm{v}}$) can be used for the aspect ambiguity sector. The basis of horizontal and vertical polarizations can be replaced by an arbitrary orthogonal polarization basis, for instance, by the basis of two opposite circular polarizations.

Consider also as examples the functions of the PSM elements after PSM representation through the diagonal matrix (1.6). Let us eliminate the influence on the signature of the fast changing parameter of the matrix (1.6), such as $\arg \mu_1$, and of its insufficiently stable parameter $\left| \mu_1 \right| = \sqrt{\sigma_1}$ being considered separately (Section 1.1). We can then use the following polarization signatures:

1. The absolute value of the ratio of PSM's eigenvalues $\left| \mu_2 / \mu_1 \right| = \sqrt{\sigma_2 / \sigma_1}$;

2. The difference of the arguments of the PSM eigenvalues $\arg \mu_2 - \arg \mu_1$.

These parameters depend on the target's type, as well as on the aspect angle. Because of possible errors in the measurement of the aspect angle, one can define the target polarization signatures only as random values for a given sector of aspect angles. So, the two-parametric probability density function (pdf) $p(\left| \mu_2 / \mu_1 \right|, \arg \mu_2 - \arg \mu_1)$ or analogous pdf $p(\left| \mu_2 / \mu_1 \right|, \left| \arg \mu_2 - \arg \mu_1 \right|)$ can define the recognition information available for the chosen signatures.

Example of Simulation. The pdf $p(\left| \mu_2 / \mu_1 \right|, \left| \arg \mu_2 - \arg \mu_1 \right|)$, $\mu_2 \leq \mu_1$ was simulated for the An-26-type aircraft and for the passive decoy missile in the aspect sector of $0°$ to $15°$ from the nose at the wavelength $\lambda = 10$ cm (Figure 3.4). The difference of probability distributions is defined in this case by the polarization transformation of the illumination signal by the propeller blades of the An-26-type aircraft and by the absence of such transformation by the Luneburg lens. However, such evident polarization differences for the aerial targets can be considered as exceptions rather than as rules.

Number of Independent Polarization Parameters for the Monostatic Radar Case. The PSM (1.3) consists of four complex elements $\dot{S}_{i,k}(i, k = 1, 2)$ or of eight scalar parameters. For the usual backscatterers with the reciprocity attribute $\dot{S}_{21} = \dot{S}_{12}^{*}$ and under conditions when one of the initial phases carries no usable information, we must consider only five independent scalar parameters of PSM. Only four such parameters will be used if one of RCS values is considered separately. Each normalized vector $\dot{X}_h$ and $\dot{X}_v$ is described by two scalar parameters. Vector $\dot{X} = \left\| \dot{X}_h \quad \dot{X}_v \right\|^{T}$ is described by four parameters: $\sqrt{\sigma_{21} / \sigma_{11}}$, $\varphi_{21} - \varphi_{11}$, $\sqrt{\sigma_{22} / \sigma_{12}}$ and $\varphi_{22} - \varphi_{12}$.

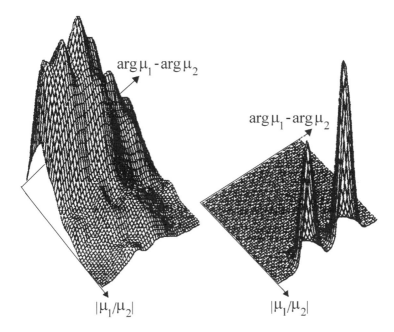

Figure 3.4 The two-parametric pdf $p(|\mu_2/\mu_1|, \arg \mu_2 - \arg \mu_1)$ simulated for the An-26-type aircraft and for the passive decoy missile in the 0° to 15° aspect sector at the wavelength $\lambda = 10$ cm.

3.3 Rotational Modulation Spectra

The rotational modulation spectra depend on:

- The type of a target;
- The wavelength;
- The aspect of a target;
- The PRF;
- The time of coherent integration.

The dependence of the spectrum on the target type determines the recognition information (Sections 3.3.1–3.3.5). One must consider this information when taking into account the other listed factors essential for recognition. Let us note that effective use of rotational modulation signatures presumes the presence of sufficient information about the propulsion engines. Since we had no reliable data about the engines, the tentative parameters were set in the backscattering model. The possibility of replacing these parameters by more reliable ones can be provided for instead of using the tentative data. For the use in recognition of the rotational modulation signatures (RMS), see also Sections 4.1.5 and 4.1.6 and Sections 4.1.9 and 4.1.10.

3.3.1 Rotational Modulation Spectra of Various Targets

Figure 3.5 shows the rotational modulation spectra of the Tu-16-type turbo-jet, the An-26-type turbo-prop aircraft, and the AH-64-type helicopter for illumination at 3-cm wavelength; the PRF is 10 kHz and the coherent integration time is about 26 ms. The aspects of the targets are 20° from the nose. All the spectra of Figure 3.5 contain the airframe line in the middle of the frequency gate and the lines of rotational modulation on both sides of it. For each stage of rotational structure, the blade frequency NF_{rot} defines the interval between the spectral lines. Here, N is the number of blades of the rotational structure stage and F_{rot} is the rotation rate. Variations of N and F_{rot} values lead to a different density of rotational modulation spectra for various types of propulsion engines.

Let us first discuss the rotational modulation spectrum of turbojet aircraft. Due to the aspect angle of 20° from the nose, the engine compressor's blades and not those of the turbine are illuminated by a radar. The rotational

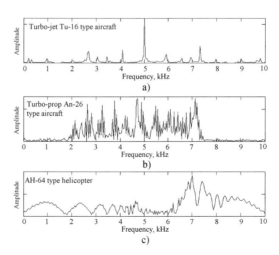

Figure 3.5 Simulated rotational modulation spectra of the (a) Tu-16-type turbojet and (b) An-26-type turbo-prop aircraft and (c) the AH-64-type helicopter for λ = 3 cm wavelength, 10 kHz PRF, and 26 ms coherent integration time. The aspects are 20° from the nose.

spectrum lines of a turbojet aircraft, the Tu-16-type here, are the most sparse of those shown in Figure 3.5. The latter are determined by the highest rotational rate F_{rot} and the largest numbers N_1 and N_2 of blades of the engine compressor's stages. Here, N_1 and N_2 are the number of blades of the first and second compressor stage. As it was experimentally justified [14, 15], the reflections from the first and second stages are essential for the formation of the rotational spectrum of a turbojet aircraft. The spectrum contains both the main blade frequencies of the first and second stages, and the combinational ones.

The rotational modulation spectrum of a turbo-prop aircraft is denser than that of a turbo-jet aircraft because the observable rotational structure (propeller) consists of fewer blades. Their length is greater and their rotational rate is less than those of a turbo-jet aircraft. The spectral lines begin to overlap, and the spectrum itself approaches a continuous one with strong asymmetry and angularity.

The rotational modulation spectrum of a helicopter is practically continuous with relatively slight asymmetry and great width. This is caused by a small number of large rotor blades rotating relatively slowly in the horizontal plane. The result of such rotor orientation is the maximum rotational modulation for almost all the aspects.

3.3.2 Rotational Modulation Spectra for Various Wavelengths

The waveband limitations of rotational modulation were discussed qualitatively in Section 1.5.1. Figure 3.6 shows the simulated spectra of rotational modulation for different wave bands: X (λ = 3 cm), C (λ = 5.25 cm), S (λ = 12.25 cm), and L (λ = 23 cm). All the spectra are given for the Tu-16-type aircraft and correspond to an aspect of 45° from the nose, PRF of 10 kHz, and coherent integration time of 26 ms. It can be seen that the signal rotational modulation caused by turbo-jet engines gradually decreases with the wavelength increase.

The signal rotational modulation caused by engines of turbo-prop aircraft and helicopters remains significant for even longer waves up to the meter ones.

3.3.3 Rotational Modulation Spectra for Various Aspects of a Target

As it was pointed out before, the rotational modulation spectrum of a helicopter does not usually depend on its aspect. But, for a turbo-jet aircraft the aspect significantly influences the rotational modulation. This influence

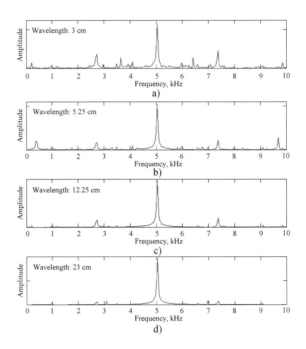

Figure 3.6 Simulated rotational modulation spectra of the Tu-16-type turbo-jet aircraft for wavelengths (a) λ = 3 cm, (b) λ = 5.25 cm, (c) λ = 12.25 cm, and (d) λ = 23 cm. The PRF is 10 kHz, coherent integration time is 26 ms, and the aspect angle is 45° from the nose.

is conditioned by two factors. First, the aspect determines whether a compressor or a turbine of the engine, or neither of them, is illuminated by the radar. Secondly, small aspect variations may cause fading of rotational modulation because of interference of signals backscattered from several engines of a multiengine aircraft.

Figure 3.7 shows the rotational modulation spectra of the Tu-16-type aircraft for aspects of 0°, 60°, and 160° from the nose. Illumination is simulated for a wavelength λ = 3 cm, with the PRF of 10 kHz, and the coherent integration time of 26 ms. Such spectra for aspects of 20° and 45° from the nose were shown in Figures 3.5(a) and 3.6(a), respectively. Rotation rates of engines were assumed here to be identical. That is not the rule. If rotation rates of engines are not identical, the angular fading transforms into temporal beatings that differ for various directions.

At target aspects of 0° and 60° from the nose the rotational modulation is conditioned by compressor blades, while at an aspect of 160° from the nose it is conditioned by turbine blades. The turbine spectrum differs then

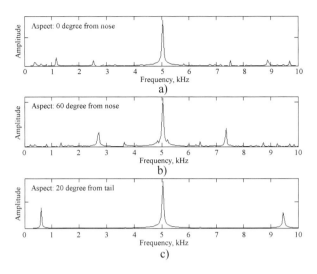

Figure 3.7 Simulated rotational modulation spectra of the Tu-16-type turbo-jet aircraft for aspect angles of (a) 0°, (b) 60°, and (c) 160° from the nose. Wavelength λ is 3 cm, the PRF is 10 kHz, and the coherent integration time is 26 ms.

from that of the compressor by having a greater blade frequency and by a smaller number of spectral lines. The great number of turbine blades leads to the screening effect and justifies frequently the "one-stage" turbine model. At that, there are the aspect sectors of very weak rotational modulation near aspects 0°, 90°, and 180° from the nose.

The fading of rotational modulation of a multiengine aircraft beyond the cited sectors can be caused by small aspect variations. This is illustrated in Figure 3.8, where the rotational spectrum of the Tu-16-type aircraft is shown for the aspects of 45°, 46°, and 47° from the nose. It can be seen that the rotational spectrum lines are changed from weak [Figure 3.8(a)] to strong [Figures 3.8(b), (c)] when the aspect variation is about 1°.

3.3.4 Rotational Modulation Spectra for Various PRFs and Coherent Integration Times

Low PRF leads to frequency aliasing of the spectral lines. In this case an unambiguous spectrum of rotational modulation can be obtained when the PRF is at least twice as high as the highest frequency of the modulation spectrum. The influence of PRF on the distortions of rotational modulation spectra is illustrated in Figure 3.9. Here, the spectra of the Tu-16-type aircraft are shown for the PRFs of 40, 6, and 2 kHz, respectively. The

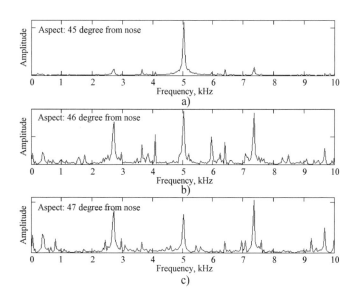

Figure 3.8 Simulated rotational modulation spectra of the Tu-16-type turbo-jet aircraft for the aspect angles of (a) 45°, (b) 46°, and (c) 47° from the nose. Wavelength λ is 3 cm, the PRF is 10 kHz, and the coherent integration time is 26 ms.

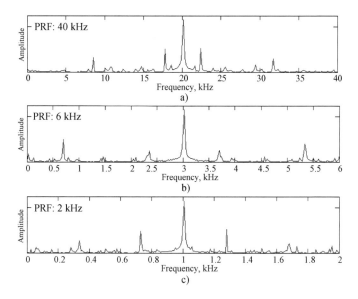

Figure 3.9 Simulated rotational modulation spectra of the Tu-16-type turbo-jet aircraft for PRF of (a) 40 kHz, (b) 6 kHz, and (c) 2 kHz. Wavelength λ is 3 cm, the coherent integration time is 26 ms, and the aspect angle is 45° from the nose.

spectrum without aliasing is obtained only for the 40-kHz PRF. For the lower PRFs (6 and 2 kHz) the spectral lines are placed at arbitrary positions due to the aliasing.

A decrease of the coherent integration interval leads to rotational modulation spectra of less detail, thus affecting their discrimination for various types of targets. Figure 3.10 illustrates the rotational modulation spectra obtained for smaller intervals of coherent accumulation of 13 and 6.5 ms. With a decrease of this interval, the spectral lines become wider.

The two factors listed above interfere with the recognition, hindering proper and precise frequency measurement. But the rotational spectrum of each target can, possibly, remain identifiable. The rotational spectrum of a turbo-jet aircraft can still be recognized from those of a turbo-prop aircraft and helicopter. Moreover, several different turbo-jet (turbo-prop, etc.) aircraft can sometimes be distinguished by their rotational spectra despite the frequency aliasing. Such possibilities require the quantitative justification (see Sections 4.1.9 and 4.1.10).

3.3.5 Comparison of Simulated Spectra with Experimental Ones

The number of available experimental publications on the subject of rotational modulation spectra is limited. R. E. Gardner's works and other works were referred to in [2]. A series of works on this subject was mentioned in Chapter 1; [3, 11–15] of this chapter are among them. The results of simulation are consistent with all the experimental data mentioned in these works. The notes about our use of provisional engine parameters were made above.

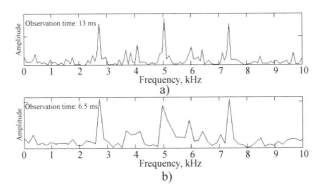

Figure 3.10 Simulated rotational modulation spectra of the Tu-16-type turbo-jet aircraft for coherent integration time of (a) 13 ms and (b) 6.5 ms. The wavelength is 3 cm, the PRF is 10 kHz, and the aspect angle is 45° from the nose.

3.4 Correlation Factors of Fluctuations Via Frequency Diversity

The frequency diversity operation is provided in some radars. We restrict the discussion here to the kind of frequency diversity where the time interval between the signals at different frequencies is small enough to assume the target is unmoved. Correlation of fluctuations at these frequencies depends on the frequency diversities and the radial size of a target.

Information Obtained from the Main Lobe of Normalized Correlation Function of Fluctuations. For the main lobe this correlation decreases with an increase of the target radial size and frequency diversity. This allows us to classify targets approximately by their radial size by estimating the fluctuation correlation for fixed frequency diversity. Normalized correlation functions obtained by simulation are shown in Figure 3.11 for the Tu-16-type and Mig-21-type aircraft and the ALCM-type missile for the aspect sector of 0° to 15° from the nose for λ = 3 cm wavelength. It can be seen that correlation of fluctuations in relation to the frequency diversity has characteristics of an antenna pattern.

Information Obtained from the Whole Normalized Correlation Function of Fluctuations. To obtain such information over a limited time interval, illumination signals at several carriers are necessary. For the small number of carriers [16], this case differs from the wideband one by the reduced information extracted from the backscattered signal. However, it is simpler to realize in unsophisticated radars.

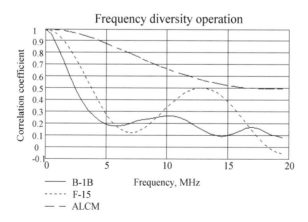

Figure 3.11   Simulated correlation coefficient of fluctuations versus frequency diversity for the Tu-16-type and Mig-21-type aircraft and the ALCM-type missile for the aspect sector of 0° to 15° from the nose. The wavelength λ is 3 cm.

References

[1]   Teti, J. G., R. P. Gorman, and W. A. Berger, "A Multifeature Decision Space Approach to Radar Target Identification," *IEEE Trans.*, AES-32, January 1996, pp. 480–487.

[2]   Nathanson, F. E., *Radar Design Principles*, New York: McGraw-Hill, 1969, pp. 171–183.

[3]   Bell, M. R., and R. A. Grubbs, "JEM Modeling and Measurement for Radar Target Identification," *IEEE Trans.*, AES-29, January 1993, pp. 73–87.

[4]   Sljusar, N. M., and N. P. Birjukov, "Backscattering Coefficient Analysis for Metallic Propeller Blades of Rectangular Form," *Applied Problems of Electrodynamics*, Leningrad: Leningrad Institute of Aviation Instrumentation, 1988, pp. 115–122 (in Russian).

[5]   Zebker, H. A, and J. J. Van Zil, "Imaging Radar Polarimetry: A Review," *Proc. IEEE*, Vol. 79, November 1991, pp. 1583–1605.

[6]   Kazakov, E. L., *Radar Recognition of Space Objects by Polarization Signatures*, Odessa: Odessa Inst. of Control and Management, 1999 (in Russian).

[7]   Shirman, Y. D., et al., "Methods of Radar Recognition and their Simulation," *Zarubeghnaya Radioelectronika—Uspehi Sovremennoi Radioelectroniki*, November 1996, Moscow, pp. 3–63 (in Russian).

[8]   Shirman, Y. D., et al., "Aerial Target Backscattering Simulation and Study of Radar Recognition, Detection and Tracking," *IEEE Int. Radar-2000*, May 2000, Washington, DC, pp. 521–526.

[9]   Barton, D. K., *Modern Radar System Analysis*, Norwood, MA: Artech House, 1988.

[10] Wilson, J. D., "Probability of Detecting Aircraft Targets," *IEEE Trans.*, AES-8, No. 6, November 1952, pp. 757–761.

[11] Tardy, I., et al., "Computational and Experimental Analysis of the Scattering by Rotating Fans," *IEEE Trans.*, AP-44, No. 10, October 1996, pp. 1414–1421.

[12] Piazza, E., "Radar Signals Analysis and Modellization in the Presence of JEM Application to Civilian ATC Radars," *IEEE AES Magazine*, January 1999, pp. 35–40.

[13] Cuomo, S., P. F. Pellegriny, and E. Piazza, "A Model Validation for the 'Jet Engine Modulation' Phenomenon," *Electronic Letters*, Vol. 30, No. 24, November 1994, pp. 2073–2074.

[14] Chernyh, M. M., et al., "Experimental Investigations of the Information Attributes of a Coherent Radar Signal," *Radiotekhnika*, March 2000, N3, pp. 47–54 (in Russian).

[15] Geyster, S. R., V. I. Kurlovich, and S. V. Shalyapin, "Experimental Studies of Spectral Portraits of Propeller-Driven Fixed-Wing and Turbo-Jet Aircraft in a Surveillance Radar with a Continuous Probing Signal," *Electromagnetic Waves and Electronic Systems*, Vol. 4, No. 1, 1999.

[16] Jouny, I., F. D. Garber, and S. Anhalt, "Classification of Radar Targets Using Synthetic Neural Networks," *IEEE Trans.*, AES-29, April 1993, pp. 336–344.

4

Review and Simulation of Recognition Algorithms' Operation

We consider at first the Bayesian (Section 4.1) and nonparametric (Section 4.2) recognition algorithms and their applications for solution of various recognition problems [1–13]. Efficiency of the recognition signatures considered previously (Chapters 2 and 3) and of their combinations can be estimated only through extensive recognition simulations. Review and simulation of Sections 4.1 and 4.2 are carried out to solve recognition problems of two kinds: recognition of target classes and recognition of target types, both with quality evaluations. The possibility of preliminary wavelet transform of data mentioned in the literature is casually discussed in Section 4.3 [14–16]. In Section 4.4 we consider neural recognition algorithms being of notable importance [17–24]. The gradient methods of training them are discussed and simulated, the quality of recognition is evaluated, and evolutionary (genetic) methods of training are also discussed. In addition, in Section 4.1.3 we consider the concept of the cooperative Recognition-Measurement Algorithm now attracting attention [25–26].

4.1 Bayesian Recognition Algorithms and Their Simulation

On the basis of the recognition quality indices introduced previously (Section 1.6.1), we consider in this chapter basic Bayesian recognition algorithms for a quasi-simple cost matrix (Section 4.1.1) and their additive variants (Section 4.1.2), which allow for the establishment of a set of efficient signatures (Sections 4.1.3–4.1.5). Such signatures as trajectory parameters and RCS are discussed in Section 4.1.3. Essential components of Bayesian additive recognition algorithms related to a set of target RPs are established and clarified in Section 4.1.4. The components of the Bayesian additive recognition algorithm related to rotational modulation and other signatures are considered in Section 4.1.5. The possibility and effectiveness of using the cpdf of RPs instead of a set of standard RPs or other signatures are considered in Section 4.1.6. Various noteworthy simulation examples of target class and type recognition are given in Sections 4.1.7 through 4.1.9 on the basis of simplified and individualized standard RPs (Section 2.2.2), cpdf of RPs (Section 2.2.2), rotational modulation spectra (Section 3.3), and other signatures. Finally, the information measures for various recognition signatures are evaluated, and the optimal target alphabets from the informational viewpoint are discussed (Section 4.1.10).

4.1.1 Basic Bayesian Algorithms of Recognition for the Quasi-simple Cost Matrix

A Posteriori Conditional Mean Risk of Recognition for the Quasi-simple Cost Matrix. The a posteriori conditional mean risk $\bar{r}(i|\mathbf{y})$ [1, 5–9] is defined under the condition that the "signal plus noise" realization $\mathbf{y}$ has been obtained. The conditional mean risk $\bar{r}(i)$ for the quasi-simple cost matrix (1.52) can be calculated according to (1.53) and can also be regarded as the result of statistical averaging of the a posteriori conditional mean risk $\bar{r}(i|\mathbf{y})$

$$\bar{r}(i) = -\sum_{k=1}^{K} r_k P_{i|k} P_k = \int_{(\mathbf{y})} r(i|\mathbf{y}) p(\mathbf{y}) d\mathbf{y} \qquad (4.1)$$

Here, r_k is a positive premium for correct recognition of the kth object, P_k is an a priori probability of appearance of the kth object, $P_{i|k}$ is a conditional probability of the decision about the presence of the ith object given the actual presence of the kth object.

The conditional probability $P_{i|k}$ depends on vector $\mathbf{y}$ of the received "signal plus noise" samples, on the chosen decision function $A_i(\mathbf{y})$ taking two values 0 and 1, and on the corresponding conditional probability density function $p_k(\mathbf{y})$ of vector $\mathbf{y}$, so that

$$P_{i|k} = \int\limits_{(\mathbf{y})} A_i(\mathbf{y}) p_k(\mathbf{y}) d\mathbf{y} \qquad (4.2)$$

Using equations (4.1) and (4.2), one obtains

$$\bar{r}(i) = \int\limits_{(\mathbf{y})} \bar{r}(i|\mathbf{y}) p(\mathbf{y}) d\mathbf{y} = -\int\limits_{(\mathbf{y})} \sum_{k=1}^{K} A_i(\mathbf{y}) p_k(\mathbf{y}) r_k P_k d\mathbf{y} \qquad (4.3)$$

The condition

$$\bar{r}(i|\mathbf{y}) p(\mathbf{y}) = -\sum_{k=1}^{K} A_i(\mathbf{y}) p_k(\mathbf{y}) r_k P_k \qquad (4.4)$$

is a sufficient condition of validity for (4.3), and it can be used to optimize recognition.

The sufficient condition of recognition optimization consists in minimization of the product $\bar{r}(i|\mathbf{y}) p(\mathbf{y})$ by means of choosing the proper number i. Since the value of $p(\mathbf{y})$ does not depend on i, it is sufficient to minimize the a posteriori conditional mean risk

$$\bar{r}(i|\mathbf{y}) = \min \qquad (4.5)$$

Basic Form of Bayesian Recognition Algorithm. To realize a single-valued recognition, only one nonzero function $A_i(\mathbf{y})$ must be chosen while minimizing the $\bar{r}(i|\mathbf{y})$. The optimal value of i (i.e., the optimal estimate $\hat{k}_{\text{opt}}$ of target class (type) k) is

$$i = \hat{k}_{\text{opt}} = \arg\max_k [p_k(\mathbf{y}) r_k P_k] \qquad (4.6)$$

Operation $\arg\max_k[\cdot]$ corresponds here to the choice of the argument k that provides the maximum value of an expression $[\cdot]$. Equation (4.6) can be considered as the general Bayesian algorithm of recognition. The generality of the simple algorithm (4.6) consists, first, in taking into account the assignment of various unequal premiums r_k for various correct decisions $k = 1, 2, \ldots, K$. So, the premium assigned for the correct recognition of a large-sized target can be higher than that of a small-sized one. The generality consists, second, in taking into account unequal a priori probabilities P_k of appearance of the targets of various classes (types).

The nearest to the optimal solution has the form

$$i_{\text{nto}} = \hat{k}_{\text{nto}} = \arg \max_{k \neq \hat{k}_{\text{opt}}} [p_k(\mathbf{y}) r_k P_k]$$

Using this solution, we can revise the reliability of optimization of the algorithm (4.6). If the values of expressions in square brackets for $k = \hat{k}_{\text{nto}}$ and $k = \hat{k}_{\text{opt}}$ are too near to each other, then both solution variants can be proposed provisionally with a simultaneous recommendation to prolong the surveillance. We consider below only the case of the single decision evaluation $k = \hat{k}_{\text{opt}}$, but the expressions being derived below may be extended to the case of evaluating two close decisions $k = \hat{k}_{\text{opt}}$ and $k = \hat{k}_{\text{nto}}$.

Variants of the Basic Bayesian Recognition Algorithm. Owing to the monotonic nature of logarithmic function, we obtain

$$i = \hat{k}_{\text{opt}} = \arg \max_k \ln[p_k(\mathbf{y}) r_k P_k] = \arg \max_k [\ln p_k(\mathbf{y}) + \ln(r_k P_k)]$$

$$(4.7)$$

One can also introduce the likelihood ratios $l_k(\mathbf{y}) = p_k(\mathbf{y})/p_n(\mathbf{y})$, where $p_n(\mathbf{y})$ is a conditional probability density function of vector $\mathbf{y}$, independent of k, corresponding to the presence of noise only. The likelihood ratio $l_k(\mathbf{y})$ widely used in the theory of signal detection can replace the conditional probability density $p_k(\mathbf{y})$ in (4.6). Then,

$$i = \hat{k}_{\text{opt}} = \arg \max_k [l_k(\mathbf{y}) r_k P_k] = \arg \max_k [\ln l_k(\mathbf{y}) + \ln(r_k P_k)]$$

$$(4.8)$$

The vector of parameters $\boldsymbol{\alpha}$ or its estimate $\hat{\boldsymbol{\alpha}}$, used in recognition, can replace, in turn, vector $\mathbf{y}$ in (4.5) and (4.6), so that

$$i = \hat{k}_{\text{opt}} = \arg \max_k \ln[p_k(\boldsymbol{\alpha}) r_k P_k] = \arg \max_k [\ln p_k(\boldsymbol{\alpha}) + \ln(r_k P_k)]$$

$$(4.9)$$

4.1.2 Additive Bayesian Recognition Algorithms

The Case of Independent Components of the Signal Parameter Vector. The conditional probability density function $p_k(\boldsymbol{\alpha}|\mathbf{y})$ of the signal parameter vector $\boldsymbol{\alpha} = \left\| \alpha_1 \quad \alpha_2 \ldots \alpha_N \right\|^{\mathrm{T}}$ can be found in this case as the product $p_k(\boldsymbol{\alpha}|\mathbf{y}) = p_k(\alpha_1|\mathbf{y}) p_k(\alpha_2|\mathbf{y}) \ldots p_k(\alpha_N|\mathbf{y})$. Its logarithm $\ln p_k(\boldsymbol{\alpha}|\mathbf{y})$ in (4.9) can be replaced then by the sum

$$\ln p_k(\boldsymbol{\alpha}|\mathbf{y}) = \ln p_k(\alpha_1|\mathbf{y}) + \ln p_k(\alpha_2|\mathbf{y}) + \ldots + \ln p_k(\alpha_N|\mathbf{y})$$
$$(4.10)$$

The Case of Independent Subrealizations of Signal. The set of the "signal plus noise" subrealizations $\mathbf{y}_1, \mathbf{y}_2, \ldots, \mathbf{y}_N$ can be received in various modes of illumination (narrowband, wideband) and at various moments of time. Such subrealizations can usually be considered as independent stochastic vectors. Then the probability density function of the "signal plus noise" vector realization $\mathbf{y} = \left\| \mathbf{y}_1 \quad \mathbf{y}_2 \quad \mathbf{y}_N \right\|^{\mathrm{T}}$ is a product of the subrealization probability densities. The logarithm of this conditional probability density $\ln p_k(\mathbf{y})$ can be introduced in the form

$$\ln p_k(\mathbf{y}) = \ln p_k(\mathbf{y}_1) + \ln p_k(\mathbf{y}_2) + \ldots + \ln p_k(\mathbf{y}_N) \qquad (4.11)$$

"Generalized" Form of Additive Recognition Algorithm. The independent parameter estimates $\hat{\boldsymbol{\alpha}}_\nu = \hat{\boldsymbol{\alpha}}_\nu(\mathbf{y}_\nu)$ are evaluated from the subrealizations $\mathbf{y}_\nu(\nu = 1, 2, \ldots N_1 \leq N)$ assumed to be independent (note that independent parameter estimates can also often be evaluated from a single subrealization). The independent logarithms of conditional likelihood ratios $\ln l_k(\mathbf{y}_k)$ are assumed to be evaluated from other independent subrealizations $\mathbf{y}_\nu(\nu = N_1 + 1, \ldots N_2 \leq N - N_1)$. Logarithms of conditional a posteriori probability density functions $\ln p_k(\mathbf{y}_k)$ in a like manner are assumed to be found from new independent subrealizations $\mathbf{y}_\nu(\nu = N_2 + 1, \ldots N)$. Generalizing (4.9) through (4.11), one obtains in this case

$$i = \hat{k}_{\mathrm{opt}} = \arg \max_k \left[\sum_{\nu=1}^{N_1} \ln p_k(\hat{\boldsymbol{\alpha}}_\nu) + \sum_{\nu=N_1+1}^{N_2} \ln l_k(\mathbf{y}_\nu) \right. \qquad (4.12)$$
$$\left. + \sum_{\nu=N_2+1}^{N} \ln p_k(\mathbf{y}_\nu) + \ln(r_k P_k) \right]$$

where only a part of the $\ln p_k(\hat{\boldsymbol{\alpha}}|\mathbf{y})$, $\ln l_k(\mathbf{y}_k)$, and $\ln p_k(\mathbf{y}_k)$ type summands can take nonzero values. Expression (4.12) emphasizes that the summands belonging to different independent signatures ν can be taken in different forms if these forms are identical for various hypotheses k, since the latter does not prevent comparing the hypotheses. Some summands of the $\ln p_k(\hat{\boldsymbol{\alpha}}_\nu)$ type can introduce only a priori information [6–9].

The Case of Interdependent Components. As is known, if there is a dependence between the random parameters α_1 and α_2, then $p(\alpha_1, \alpha_2) = p(\alpha_1)p(\alpha_2|\alpha_1)$. Accordingly, (4.10) takes the form

$$\ln p_k(\boldsymbol{\alpha}|\mathbf{y}) = \ln p_k(\alpha_1|\mathbf{y}) + \ln p_k(\alpha_2|\mathbf{y}, \alpha_1) + \ldots$$
$$+ \ln p_k(\alpha_N|\mathbf{y}, \alpha_1, \ldots \alpha_{N-1})$$

Additivity of algorithms in the simplest sense [i.e., according to (4.10) through (4.12)] becomes nonoptimal. Using the simulation methods, however, one can be assured that in many cases the deviation from the optimal algorithm is not very significant. In cases when this deviation is significant, one can reduce its importance by increasing the interval between observations or combining the interdependent summands of expression (4.12) into enlarged ones. These algorithms are no worse, evidently, than the widely used voting algorithms described in Section 4.2.

4.1.3 Components of Additive Bayesian Recognition Algorithms Related to the Target Trajectory and RCS

Trajectory components of the additive algorithm (4.12) can be chosen in the form $\ln p_k(\hat{\boldsymbol{\alpha}}_\nu)$, $\nu = \nu_{tr}$ for various hypotheses k. These components take into account the altitudes, vector velocities, and accelerations of flying targets. The trajectory information taken alone is usually insufficient for correct recognition due to overlapping of parameter distribution regions for various target classes, but it can be useful for recognition together with other kinds of information about the target.

The Simplest Trajectory Parameters in the Case of Target Class Recognition. Figure 4.1 shows tentatively the distribution regions of the two-dimensional parameter $\boldsymbol{\alpha} = \|\alpha_1 \quad \alpha_2\|^T$ including the velocity $V = \alpha_1$ and altitude $H = \alpha_2$ for some classes of aerial targets. The distribution regions are presented for the classes: (1) large-sized aircraft, (2) medium-sized aircraft, (3) missile, [(3′) cruise missile], and (4) helicopter.

Figure 4.2 shows tentatively the distribution regions of the two-dimensional parameter $\boldsymbol{\alpha}$ including the velocity $V = \alpha_1$ and altitude $H = \alpha_2$ for 11 types of aerial targets.

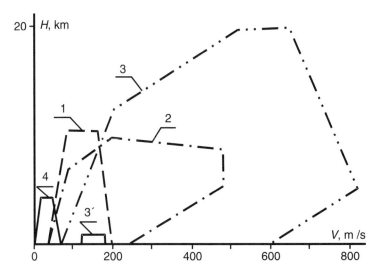

Figure 4.1 Tentative distribution regions of the "velocity *V*-altitude *H*" 2D parameter for various classes of aerial targets: (1) large-sized aircraft, (2) medium-sized aircraft, (3) missile, (3') cruise missile, and (4) helicopter.

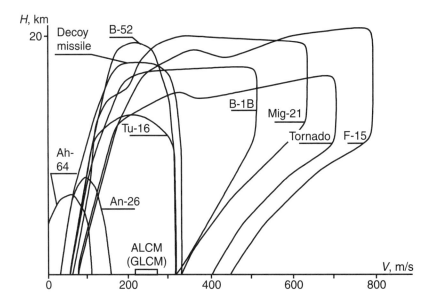

Figure 4.2 Tentative distribution regions of the "velocity *V*-altitude *H*" 2D parameter for various types of aerial targets: Tu-16, B-52, B-1B, Mig-21, F-15, Tornado, An-26, AH-64, ALCM, GLCM, and passive decoy.

Formalization of the Parameter Distribution Regions.   To formalize the distribution regions we can use a polygon or a matrix approximation.

The polygon approximation is introduced assuming that the a priori probabilities of the parameter estimates $\hat{\boldsymbol{\alpha}}$ are distributed uniformly within these polygon regions, so that $\ln p_k(\hat{\boldsymbol{\alpha}}) = \ln p_k$, if $\mathbf{A}_k\hat{\boldsymbol{\alpha}} + \mathbf{B}_k \geq 0$, and $\ln p_k(\hat{\boldsymbol{\alpha}}) = -\infty$ or $p_k = 0$, if $\mathbf{A}_k\hat{\boldsymbol{\alpha}} + \mathbf{B}_k < 0$. Correspondingly, the summation in (4.12) must be carried out for the nonzero a priori probability densities $p_k \neq 0$ only. The value of a probability density p_k is defined by the inverse value of a kth polygon area. The vector-matrix inequalities must be understood as a set of scalar inequalities, defining position of the point $\hat{\boldsymbol{\alpha}}$ inside or on the border of the kth polygon. In the case of three or more components of the parameter vector $\boldsymbol{\alpha}$, the results for the polygon regions can be generalized to polyhedron regions. In the case when the measurement errors are significant, the size of the polygon (polyhedron) regions has to be correspondingly enlarged.

The matrix approximation is described by K matrices with scalar multipliers instead of K sets of matrix inequalities (see above). This approximation is introduced on the assumption of replacing a continuous description of the distribution region borders by discrete distributions within a rectangular part of the velocity-altitude (V-H) plane embracing all the distribution regions. As it was above, the a priori probability of the parameter vector hit within each distribution region is assumed to be spread uniformly. The elements of the kth matrix lying inside or on the border of the kth region become units, and the other elements lying outside the region become zeros. The matrix of units and zeros for each kth object is multiplied by $P_k = \mu_k^{-1}$. Here, μ_k is the ratio of the number of the kth matrix elements equal to unity to the whole number of its elements.

Additional Possibilities of Using the Trajectory Information in the Case of Precise Target Tracking.   In the case of target tracking with sufficiently frequent illumination, there are additional possibilities of using trajectory information in more detail. One can use the vertical velocity component and various accelerations. Especially interesting is the possibility of measuring the heading and heading rate of a target more precisely by using the extended Kalman filter, which allows the more efficient use of RP information.

Some interesting suggestions arise about verification of the "target type–target aspect" complex hypothesis; see, for instance, [25, 26]. Instead of choosing between K hypotheses about K target types in the aspect sector (aspect uncertainty sector) $\Delta\beta$, we can choose between PK complex hypotheses about the target to be of kth type and to be in the aspect sector

$p \Delta \beta / P(k = 1, 2, \ldots, K; p = 1, 2, \ldots, P)$. The hope is that this technique will also improve the quality of target tracking in complex situations.

The RCS Component. This component of the additive algorithm (4.12) can be chosen for various hypotheses k in the form $\ln p_k(\hat{\boldsymbol{\alpha}}_\nu)$, where $\nu = \nu_{\mathrm{RCS}}$ and $\hat{\boldsymbol{\alpha}}_\nu = \| \hat{\sigma}_1 \quad \hat{\sigma}_2 \ldots \|^{\mathrm{T}}$. In the case of the approximately normal RCS distribution, $\hat{\boldsymbol{\alpha}}_\nu \approx \hat{\alpha}_\nu = \dfrac{1}{n} \sum\limits_{\mu=1}^{n} \sigma_\mu$. The components σ_μ can take into account the introduction of RCS in linear or logarithmic scale, at single or several carrier frequencies, and for the case of narrowband and wideband signal use. In the case of a narrowband signal, n is only the number of independent measurements. In the case of a wideband signal, each value of σ_μ is obtained on its turn as a sum of the RCSs of resolved target elements. Multiplicative distortions observed in the linear scale become additive in the logarithmic scale.

Example of "Radial Extent-RCS" Distribution Regions. In addition to the "velocity-altitude" distribution regions of Figures 4.1 and 4.2, we show in Figure 4.3 the "radial extent-RCS" distribution regions for the following target classes: (1) large-sized aircraft, (2) medium-sized aircraft, (3) missile, (4) small-sized passive decoy with the RCS greatly increased due to the

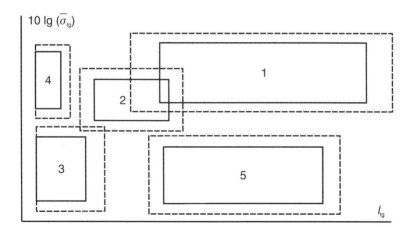

Figure 4.3 Tentative distribution regions of the two-dimensional parameter "radial extent-RCS" for the aerial target classes: (1) large-sized aircraft, (2) medium-sized aircraft, (3) missile, (4) passive decoy, and (5) aircraft with significantly decreased RCS.

Luneburg lens or corner reflector installed on it, and (5) aircraft with significantly decreased RCS.

4.1.4 Component of Additive Bayesian Recognition Algorithms Related to Correlation Processing of Range Profiles

The RP component $\nu = \nu_{RP}$ of the additive algorithm (4.12) for various hypotheses k can be chosen in the form of the logarithm of a cpdf, multidimensional in this case, $\ln p_k(\boldsymbol{\alpha}_\nu | \mathbf{y}_\nu) \approx \ln p_k(\hat{\boldsymbol{\alpha}}_\nu)$, where $\hat{\boldsymbol{\alpha}}_\nu = \hat{\boldsymbol{\alpha}}_\nu(\mathbf{y}_\nu)$, $\nu = \nu_{RP}$. To realize this algorithm directly, it is necessary to estimate the multidimensional conditional probability density function (multidimensional cpdf) $p_k(\boldsymbol{\alpha}_\nu)$ for various values of k. Large a priori RP statistics are therefore necessary. Hence, the correlation algorithms with some standard profiles described above (Section 2.2.2) were used at first in the theoretical and experimental investigations. These algorithms are the approximations of the additive recognition algorithm (4.12), as it will be shown below. They approximate the algorithm based on the multidimensional cpdf $p_k(\boldsymbol{\alpha}_\nu)$ for various k (Section 4.1.6) as the number of standard RPs increases.

The Model of Entirely Known RP for Each Object. It is considered here as an intermediate step toward the model of a known set of standard RPs for each object [6, 9].

Let us introduce the expected single RP $\mathbf{X}$ of a single object being sampled and normalized to unit energy, so that $\mathbf{X} = \left\| X_m \right\|$, $m = 1, \ldots, M$ and $\sum_{m=1}^{M} X_m^2 = 1$. Let us consider this RP as a burst of expected complex samples $\dot{\mathbf{X}}(\boldsymbol{\beta}, b) = b \left\| X_m \exp(j\beta_m) \right\|$ with unknown phases β_m and a common multiplier b. This burst of signal samples is superimposed on the burst $\dot{\mathbf{N}} = \left\| \dot{N}_m \right\|$ of independent samples of noise, each with variance σ^2. The probability density function of noise samples is

$$p(\operatorname{Re}\dot{\mathbf{N}}, \operatorname{Im}\dot{\mathbf{N}}) = p_N(\dot{\mathbf{N}}) = (2\pi\sigma^2)^{-M} \exp(-\left|\dot{\mathbf{N}}\right|^2 / 2\sigma^2)$$

The superposition $\dot{\mathbf{Y}} = \dot{\mathbf{N}} + \dot{\mathbf{X}}(\boldsymbol{\beta}, b)$ for given β_m and b has the probability density function

$$p_{SN}(\dot{\mathbf{Y}}) = p_N[\dot{\mathbf{Y}} - \dot{\mathbf{X}}(\boldsymbol{\beta}, b)]$$

Averaging of $p_{SN}(\dot{\mathbf{Y}})$ by phases and amplitude leads to difficult calculations. Let us find, therefore, the maximum likelihood estimates for all β_m and b from the equations

$$\frac{\partial}{\partial \beta_m} \ln p_N[\dot{\mathbf{Y}} - \dot{\mathbf{X}}(\boldsymbol{\beta}, b)] = 0, \qquad \frac{\partial}{\partial b} \ln p_N[\dot{\mathbf{Y}} - \dot{\mathbf{X}}(\boldsymbol{\beta}, b)] = 0$$

$$(4.13)$$

One can obtain the estimates $\hat{\beta}_m = \arg \dot{Y}_m$ from the equation

$$\partial \left| \dot{\mathbf{Y}} - \dot{\mathbf{X}}(\boldsymbol{\beta}, b) \right|^2 / \partial \beta_m = \partial \left[(\dot{Y}_m - b X_m e^{j\beta_m})(\dot{Y}_m^* - b X_m e^{-j\beta_m}) \right] / \partial \beta_m = 0$$

being equivalent to the first of equation (4.13). The estimate

$$\hat{b} = \sum_m \left| \dot{Y}_m \right| X_m \Big/ \sum_m X_m^2 = \sum_m \left| \dot{Y}_m \right| X_m$$

can be obtained from the second of equations (4.13) and the normalization condition $\sum_m X_m^2 = 1$. After replacing the $p_{SN}(\dot{\mathbf{Y}})$ by its estimate, we can evaluate the logarithm of the likelihood ratio

$$\ln \hat{l}(\dot{\mathbf{Y}}) = \ln[\hat{p}_{SN}(\dot{\mathbf{Y}}) / p_N(\dot{\mathbf{Y}})]$$

$$= -\frac{1}{2\sigma^2} \sum_{m=1}^{M} \left[(\dot{Y}_m - \hat{b} X_m e^{j\beta_m})(\dot{Y}_m - \hat{b} X_m e^{j\beta_m})^* - \left| \dot{Y}_m \right|^2 \right]$$

$$= Z_\Sigma^2 / 2\sigma^2 \qquad (4.14)$$

that is to replace the $\ln l(\dot{\mathbf{Y}})$ entered in (4.12).

The logarithm of the likelihood ratio (4.14) is a monotonically increasing function of the correlation sum

$$Z_\Sigma = \sum_m \left| \dot{Y}_m \right| X_m \qquad (4.15)$$

that was previously given by (2.4) without statistical grounding.

In the case of $k = 1, 2, \ldots, K$ various objects with the single standard RP $\mathbf{X} = \mathbf{X}_k = \left\| X_{mk} \right\|$ for each object, the selection of maximum value of correlation sum $Z_\Sigma = Z_{\Sigma k}$ permits the evaluation of the class or type of object.

The Model of a Known Set of Standard RPs for Each Object. It is used because the orientation of an aerial target can be evaluated only approximately and is defined by a comparatively broad aspect sector. An arbitrary object of kth class or type is described then by several $\gamma = 1, 2, \ldots, \Gamma$ standard RPs $\mathbf{X}_{\gamma|k} = \|X_{m,\gamma|k}\|$. Each γth standard RP of the kth object is characterized with a priori conditional probabilities $P_{\gamma|k}$, the sum of which $\sum_{\gamma=1}^{\Gamma} P_{\gamma|k}$ is equal to unity. Expression (4.14) takes the form

$$\ln l_k(\mathbf{Y}) = \ln\left[\sum_{\gamma=1}^{\Gamma} P_{\gamma|k} \exp(Z_{\Sigma\gamma|k}^2 / 2\sigma^2)\right] \qquad (4.16)$$

where $Z_{\Sigma\gamma|k}$ is the correlation sum (4.15) for $X_m = X_{m,\gamma|k}$.

The Case of Imprecisely Known Range of a Target. It is typical when the wideband recognition signals are radiated immediately after radiation of narrowband signals. The imprecisely known target range can hamper the operation of the presented algorithms. But if the potential SNR is sufficiently great, the maximum probable range can be evaluated directly from the raw recognition data. So, the correlation sum $Z_{\Sigma\gamma|k}$ in (4.16) can be replaced by

$$Z_{\Sigma\gamma|k} = \max_{\mu} \sum_{m} |\dot{Y}_m| X_{(m-\mu)\gamma|k} \qquad (4.17)$$

where the optimal value of μ elaborates the estimate of target range and matches the received and standard RPs. Operation (4.17) can be considered as a discrete linear filtration of the received RP samples $|Y_m|$ by the filter matched to the standard RP $\mathbf{X}_{\gamma|k}$.

Since the exact time of the echo signal arrival is arbitrary, the sampling moments for the RP and for the standard RP do not coincide, in the general case. To avoid large errors in comparison of the similar hypotheses, it is desirable that the signal sampling frequency in (4.17) be several times greater than its bandwidth.

Stricter asymptotic grounding of (4.17) is analogous to the simplification of (4.16) considered below in the case of asymptotically large SNR.

The Case of Asymptotically Large SNR for a Known Set of RPs. The exponential form of summands (4.16) leads to domination, over the rest of summands, of the one corresponding to the maximum value of the product $P_{\max|k} \exp\left(\dfrac{Z_{\max|k}^2}{2\sigma^2}\right)$, which is calculated for some number $\gamma = \gamma_{\max}$ of standard RP. After simple algebraic transform of (4.16), one obtains

$$\ln l_k(\mathbf{Y}) \approx \ln\left\{\left[P_{\max|k} \exp\left(\frac{Z^2_{\max|k}}{2\sigma^2}\right)\right]\right.$$

$$\left.\cdot\left[1 + \sum_{\gamma \neq \gamma_{\max}} \frac{P_{\gamma|k}}{P_{\max|k}} \exp\left(-\frac{Z^2_{\max|k} - Z^2_{\gamma|k}}{2\sigma^2}\right)\right]\right\}$$

Computing the logarithm of multipliers' product and using the approximate equation $\ln(1 + x) \approx x$ that is correct for $|x| \ll 1$, we have

$$\ln l_k(\mathbf{Y}) \approx \ln P_{\max|k} + \frac{Z^2_{\max|k}}{2\sigma^2} + \sum_{\gamma \neq \gamma_{\max}} \frac{P_{\gamma|k}}{P_{\max|k}} \exp\left(-\frac{Z^2_{\max|k} - Z^2_{\gamma|k}}{2\sigma^2}\right)$$

For very large SNR the sum of the first two summands dominates over the others, and we obtain

$$\ln l_k(\mathbf{Y}) \approx \max_{\gamma}\left[\ln P_{\gamma|k} + \frac{Z^2_{\gamma|k}}{2\sigma^2}\right] \tag{4.18}$$

Equation (4.18) can be used independently or as a part of (4.12). The summands $\ln P_{\gamma|k}$ were absent in the heuristic formulae of Section 2.2.2. They can be significant only if the values $P_{\gamma|k}$ are unequal for the different numbers γ and k.

4.1.5 Components of Additive Bayesian Recognition Algorithms Related to Correlation Processing of the RMS and Other Signatures

The use of rotational modulation spectra (Section 3.3) can be carried out in various forms. One such form envisages the correlation processing of a spectrum part under the condition of fixed and stable PRF. The use of a PRF that avoids spectrum overlapping is desirable, but escapable in this case.

Let us consider the sampled amplitude-frequency spectrum $G(F_k)$ ($k = 1, 2, \ldots, K$) of a burst of pulses having constant PRF and subjected to rotational modulation (Sections 1.5 and 3.3). The set $\mathbf{G}$ of such spectrum samples $G(F_k)$ within the frequency interval $F_D \leq F_k < F_D + F_{pr}$, where F_D is the target doppler frequency and F_{pr} is PRF, can be considered as the RMS analogous to the RP. Like the RP, the RMS can be normalized. Correlation processing can also be used. Analogous to the simplest and individualized standard RPs (Section 2.2.2) the simplest and individualized standard RMSs may be introduced.

In the simplest case, the cooperative use of rotational modulation and RPs consists of comparison of two or many successive RPs to reveal the influence of the rotational modulation sources. As it was shown in Figure 2.6, the successive RPs of a helicopter are slightly correlated, and this can be used as an additional signature. Such a signature was observed first in the 1980 experiment on a radar heightfinder operated in the high-frequency part of S-band (Section 2.2.5), but it can also be used in a very broad frequency band.

As was shown in Figures 2.2 through 2.6 and 2.19, the positions of the aircraft rotating parts are also observable on an RP at the wavelengths $\lambda < 10$ cm as an additional signature. Increasing a number of successive RPs allows us to obtain range-frequency signatures (Figure 2.19) combining RP and rotational information. Two-dimensional correlation processing with matching of the type (4.17) in range and in frequency can be used for this purpose. Obtaining the range-frequency signature presumes that the duration of the coherent pulse burst is sufficient for an acceptable spectral resolution of rotational modulation components and insufficient for spectral resolution of the target body cross-range elements. All the signatures considered here, like the RP, can be grounded theoretically if one considers an independent subrealization $\mathbf{y}_\nu$ of additive algorithm (4.12) including more than one RP.

Use of ISAR procedures and two-dimensional images presupposes the pulse burst duration to be sufficient for resolution of the cross-range elements of a target body (Section 2.5). The amplitude information necessary for recognition can be used in a 1-bit (binary) [2] or multibit digital form. Two-dimensional correlation processing requires us to ensure the matching in three parameters: range, cross range, and aspect angle. Together with correlation processing, the neural one (Section 4.4) can be carried out, but the recognition algorithm operation in this case remains insufficiently studied.

Let us now discuss the use of range-polarization and polarization profiles. Since these signatures (Sections 2.3.1 and 3.2.2) have a form analogous to the RP form, they can be subjected to correlation processing, coherent or noncoherent, analogous to that described above. Specifics of coherent correlation processing (2.8), which can be reasoned strictly, consist of accounting for the complex character of correlation sums.

4.1.6   Use of cpdf Instead of Sets of RPs, RMSs, or Other Signatures

Evaluation of the cpdf of RPs.   An arbitrary pdf of a scalar random variable x can be evaluated by smoothing its histogram or by using the method of

Parzen window functions [2]. It is convenient to employ the Gaussian window function

$$\varphi(u) = \frac{1}{\sqrt{2\pi}} e^{-0.5u^2}$$

We suppose that a great number N of experimental or simulated values Y_1, Y_2, ..., Y_N of the scalar random variable Y is obtained. The pdf approximation by Parzen windows takes the form

$$p(Y) \approx \frac{1}{NA} \sum_{\nu=1}^{N} \varphi\left(\frac{Y - Y_\nu}{A}\right)$$

where A is a supplementary scale parameter that is defined from the pdf normalization condition

$$\int_{-\infty}^{\infty} p(Y)dY = 1$$

The described method of evaluation of the pdf $p(Y)$ is also applicable to evaluate the cpdf $p_k(Y)$ introduced under the condition that a target of the kth ($k = 1, 2, \ldots, K$) class (type) is present.

Use of the cpdf of RPs or RMSs Instead of a Set of Standard RPs or RMSs. Instead of introducing various normalized standard RPs or RMSs, one can introduce the cpdf of the individual samples of RP or RMS under the assumption of their independence. The summand $\ln p_k(\mathbf{y}_\nu)$ of (4.12), where ν is the number of subrealizations, can be replaced by

$$\ln p_k(\mathbf{y}_\nu) = \ln \prod_{m=1}^{M} p_{m|k}(Y_{m\nu}) = \sum_{m=1}^{M} \ln p_{m|k}(Y_{m\nu})$$

Here, $p_{m|k}(Y_{m\nu})$ is the cpdf of the mth sample of the kth RP or RMS received by means of the νth signal subrealization for an uncertainty sector of target orientation. If the SNR is high, then the cpdf obtained without noise can approximate the cpdf obtained in the presence of noise. But the cpdf approximation for the SNR of 20 to 30 dB proved to be more suitable in simulation than that for the SNR tending to infinity (see Section 4.1.8).

Values of cpdf $p_{m|k}(Y)$ or lcpdf $\ln p_{m|k}(Y)$ must be stored in a computer's memory and used after reception of all the samples $Y_{m|\nu}$.

The Case of Imprecisely Known Range or Doppler Frequency of a Target. Using the RPs or RMSs, we have in this case

$$\ln p_k(\mathbf{y}_\nu) = \ln\left[\sum_\mu P_\mu \prod_{m=1}^M p_{(m-\mu)|k}(Y_{(m-\mu)\nu})\right]$$

where P_μ is the a priori probability of an error in range or doppler frequency equal to μ sampling intervals, so that $\sum_\mu P_\mu = 1$. The distribution of values P_μ corresponds usually to the normal pdf with a variance depending on the accuracy of range or doppler frequency measurement.

If the potential SNR is high, the latter equation takes the filtration-like form

$$\ln p_k(\mathbf{y}_\nu) = \max_\mu\left[\ln P_\mu + \sum_{m=1}^M \ln p_{(m-\mu)|k}(Y_{(m-\mu)\nu})\right] \qquad (4.19)$$

where the chosen value of μ maximizes (4.19).

The equations obtained above can be employed not only for the use of RP, but also for RMS. The optimal value of μ in (4.19) then optimizes the estimated doppler frequency or range of a target.

4.1.7 Simulation of Target Class Recognition Using the Simplest Standard RPs and Other Signatures

Target Class Recognition Using the RPs, RCSs, and Trajectory Signatures. Radar recognition of four target classes was simulated using a chirp signal of 80-MHz bandwidth in the high-frequency part of S-band [7, 9].

The recognized classes were: (1) large-sized aircraft (Tu-16, B-52, and B-1B), (2) medium-sized aircraft (Mig-21, F-15, and Tornado), (3) missile (ALCM and GLCM types), and (4) passive decoy.

Three signature sets (SSs) were considered:

- SS1: the correlation sums for $\Gamma = 3$ simplest standard RPs on a class (12 on the whole) together with the wideband RCS. The simplest standard RPs were obtained for target classes (without their individualization to target types) for aspect sectors of $20°$ on the basis of several tens $T > \Gamma$ of simulated training RPs on a sector (Section 2.2);

- SS2: the set SS1 + RCS estimated by 10 to 15 narrowband illuminations;

- SS3: the set SS2 + trajectory signatures (the velocity and altitude without specification for target types).

Together with accounting for the RCS fluctuations, an a priori error of RCS measurement (due to instabilities of propagation conditions and parameters of a transmitter and receiver) was considered with variance of 3 dB.

The factor of imprecisely known range of a target was considered here also. The operation of the RP coarse centering was carried out, as it was in the 1985–1987 experiment. To reduce computational expenses, the coarse centering was provided on the basis of auxiliary rectangular RPs of various durations corresponding to various hypotheses of target classes. The sampling frequency was chosen two times higher than the minimal frequency corresponding to the sampling theorem of Kotelnikov-Shannon. Unlike in the experiment, the training in simulation was carried out on the basis of RPs formed in the absence of noise.

For testing target class recognition, we simulated equiprobable appearance of various class objects. Figure 4.4 shows the simulated probabilities of errors in target class recognition versus potential SNR in dB for single ($N = 1$) target illumination by a wideband signal of 80-MHz bandwidth and for the sets of signatures SS1, SS2, SS3. The results of simulation for the set of signatures SS1 correspond in the whole to results of the 1985–1987 experiment (Section 2.2.5). Unlike this experiment, the training of the recognition algorithm was carried out in simulation without the noise background.

Corresponding simulated probabilities of errors versus potential SNR in dB for several ($N \geq 1$) target illuminations by a wideband chirp signal of 80-MHz bandwidth and for the set of signatures SS1 are shown in Figure 4.5. It is seen that recognition quality is improved when the number of illuminations increases.

We see that the potential SNR for recognition, which is about 20 dB or even more in our case, must be higher than for detection, which is 13 to 15 dB. The use of only several tens of training RPs, of three standard RPs on a type, and of coarse RP centering (all that was conditioned by

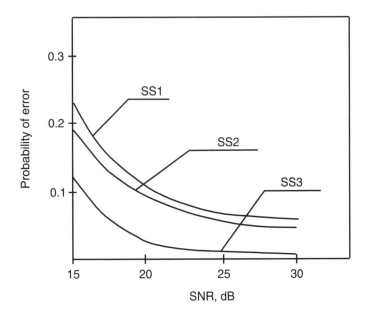

Figure 4.4 Simulated probabilities of errors in target class recognition, with equiprobable object appearance, versus potential SNR for single ($N = 1$) target illumination by a chirp signal of 80-MHz bandwidth, for the sets of signatures SS1, SS2, SS3 (all with the simplest standard RPs and their coarse centering).

limited experimental and computational possibilities in the 1980s) was then applied to succeeding simulations. Increases in the number of training and standard RPs, in quality of their centering, in potential SNR, in number of illuminations, and in signal bandwidth can improve the recognition quality.

Specification of Target Class Recognition Using the RPs. The following simulation was provided for recognition of three classes of targets by means of several illuminations with a chirp signal of 80-MHz bandwidth in the high-frequency part of S-band. The number T of training RPs was increased up to 100, but the number Γ of standard RPs for each class in the 20° aspect sector remained equal to 3. The appearance of targets of various classes was assumed to be equiprobable (P_k = const); the importance of recognition of all the targets was also assumed to be equivalent (r_k = const). The summand $\ln(r_k P_k)$ of (4.12) common here for all the recognized classes was neglected, unlike in the previous case (see below). The classes to be recognized were the same as above, but without the passive decoy. The standard RPs remained not individualized according to the target types. Coarse RP centering was

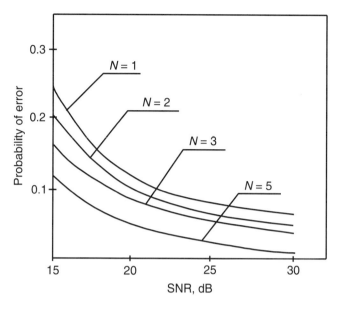

Figure 4.5 Simulated probabilities of errors in the target class recognition, with equiprobable object appearance, versus potential SNR for several ($N \geq 1$) target illuminations by a chirp signal of 80-MHz bandwidth for the set of signatures SS1 with simplest standard RPs and their coarse centering.

provided on the basis of auxiliary rectangular RPs (with the extent of 27 samples for the hypothesis of a large-sized aircraft, of 15 samples for the hypothesis of a medium-sized aircraft, and of 7 samples for the hypothesis of a missile).

In Table 4.1 we specify the errors showing the numbers of various error and correct decisions made after 100 presentations of each of the RP groups including one ($N = 1$), two ($N = 2$), and three ($N = 3$) independent RPs each. The RPs were simulated on a noise background in the 20° aspect sector. The numbers of correct and error decisions in Table 4.1 are random, to a certain degree, due to randomness of the noise and RP realizations. Due to this randomness, some results of Table 4.1 for the larger values of N and SNR are worse than those for the smaller ones. But, in general, as above, the recognition quality increases with increasing the SNR and N.

It can be seen that the worst conditions of recognition are for the medium-sized targets, which can be misrecognized both as large- and small-sized ones. Errors' redistribution can be achieved by restoration of the summand $\ln(r_k P_k)$ of (4.12) and increasing the value r_k for medium-sized targets. We used such redistribution in the 1985–1987 experiment and in

Table 4.1
Numbers of Correct and Error Decisions About the Target Class Made Using Simplest Standard RPs and Coarse RP Centering After 100 Presentations of the RPs' Groups of Each Class

Presented Targets			Number N of Independent RPs in a Group								
			$N = 1$			$N = 2$			$N = 3$		
			Number $i = \hat{k}$ of Decisions about the Target Class			Number $i = \hat{k}$ of Decisions about the Target Class			Number $i = \hat{k}$ of Decisions about the Target Class		
Class k	Type	SNR dB	$i = 1$	$i = 2$	$i = 3$	$i = 1$	$i = 2$	$i = 3$	$i = 1$	$i = 2$	$i = 3$
1	Tu-16	18	91	3	6	98	1	1	99	1	0
		21	85	4	11	95	0	5	98	0	2
		24	96	3	1	100	0	0	100	0	0
	B-52	18	87	10	3	88	10	2	91	7	2
		21	78	19	3	87	12	1	88	12	0
		24	79	10	11	90	7	3	95	5	0
	B-1B	18	77	7	16	82	4	14	82	3	15
		21	81	6	13	82	4	14	82	4	14
		24	85	9	6	94	5	1	94	5	1
2	Mig-21	18	13	55	32	4	63	33	0	81	19
		21	6	70	24	0	76	26	0	89	11
		24	3	70	27	0	85	15	0	89	11
	F-15	18	14	53	32	6	62	32	3	78	19
		21	8	64	28	2	73	25	0	75	25
		24	5	71	24	0	89	11	0	86	14
	Tornado	18	115	67	18	12	82	6	8	86	6
		21	7	72	21	1	89	10	2	91	7
		24	3	85	12	0	94	6	0	97	3
3	ALCM	18	11	10	79	1	5	94	1	3	96
		21	4	7	89	0	1	99	0	1	99
		24	1	4	95	0	0	100	0	1	99
	GLCM	18	7	2	91	1	0	99	0	0	100
		21	4	0	96	1	0	99	0	0	100
		24	1	1	98	0	0	100	0	0	100

the simulation described above. It is interesting also that the large-sized aircraft can be misrecognized not only as a medium-sized aircraft, but also as a missile. The latter is because the RPs of large targets can become peaked due to the interference of unresolved target elements. On the whole, the results of Table 4.1 are somewhat better than those of Figures 4.4 and 4.5 due more to the increased number of training RPs than to the decreased number of hypotheses to be compared.

4.1.8 Simulation of Target Type and Class Recognition Using Individualized Standard RPs and cpdf of RPs

Target Type Recognition Using the Individualized Standard RPs and cpdf of RPs. The recognition quality can be significantly improved if:

• The number of training RPs is increased from several tens to about several hundreds on a type;

• The number of standard RPs is also increased allowing the use of individualized standard RPs or, in the limit, the cpdf of the RPs;

• Instead of coarse RP centering based on auxiliary rectangular RPs, the filtration-like one [(4.17) and (4.19)] is used. For wideband radar with precision tracking in range, the centering problem becomes easier.

Figure 4.6 shows the simulated probabilities of errors in target type recognition, with equiprobable appearance of 11 objects, versus potential SNR in dB for various recognition algorithms:

• Correlation approximations of the additive Bayesian algorithm with various numbers Γ = 1, 3, 5, 10 of individualized standard RPs on a type (11, 33, 55, 110 standard RPs on a sector in the whole), with filtration-like centering (4.17) and with the minimal Kotelnikov-Shannon sampling frequency;

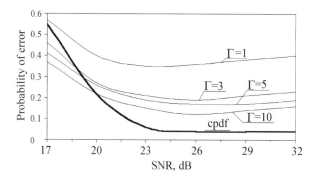

Figure 4.6 Simulated probabilities of errors in the type recognition of 11 objects versus potential SNR for a single ($N = 1$) target illumination by an 80-MHz chirp signal. The Γ = 1, Γ = 3, Γ = 5, and Γ = 10 curves correspond to the correlation processing with various numbers Γ of individualized standard RPs on a type in an aspect sector 0°–20° and filtration-like RP centering. The cpdf curve corresponds to cpdf processing and exactly known range.

- Additive Bayesian algorithm with direct evaluation of conditional probability density functions of the RPs (cpdf processing).

In case of cpdf, the target range was assumed to be exactly known. For a radar with narrow- and wideband signals, such an assumption makes the recognition performance somewhat higher compared to filtration-like RP centering (4.19). For a tracking radar with only wideband signals, such an assumption can be justified, though the theory and technique of such tracking are not yet developed sufficiently. The estimate of the RP cpdf was obtained using the Parzen window (Section 4.1.6). Training sets of T = 300 RPs for each target (3300 RPs in the whole on a sector) were used, and the training was carried out in presence of noise under the assumption of potential SNR to be arbitrarily chosen for each RP from 20 to 30 dB. A wideband chirp signal in S-band with 80-MHz bandwidth was simulated to produce the RPs in the nose-on aspect sector of 0° to 20°. In the processes of training and testing the target pitch angle was changed monotonically from −1° to 9° and the target roll angle was changed from −5° to 5°. The sampling frequency was chosen equal to the minimal frequency corresponding to the sampling theorem.

At the test stage it was supposed that each decision was made using the only RP ($N = 1$). The recognition performance was tested here using the training set of RPs superimposed on the realizations of noise different from realizations used in the training. The simulated objects were Tu-16, B-52, B-1B, An-26, Mig-21, F-15, and Tornado aircraft, an AH-64 helicopter, missiles of ALCM and GLCM types, and a passive decoy. As was assumed here, the decoy had only one Luneburg lens, so that its RP could be distinguished from that of a missile without using any amplitude information.

Figure 4.7 shows the simulated probabilities of errors in target type recognition for single target illumination versus bandwidth of a chirp signal for the case of equiprobable appearance of 11 objects (curves 1 and 2) and of six objects (curves 3 and 4). Curves 1 and 3 correspond to the correlation algorithm of recognition with $\Gamma = 3$ individualized standard profiles. Curves 2 and 4 correspond to cpdf processing. The radar band, aspect sector, and training set were the same as for Figure 4.6. Simulation was carried out for the SNR of 25 dB. Filtration-like centering (4.17) was used in correlation processing (curves 1 and 3). Curves 2 and 4 corresponding to cpdf processing were obtained for exactly known target range. These results show us that the signal bandwidth must be increased if the number of objects to be recognized is increased.

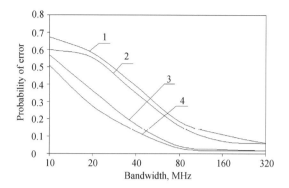

Figure 4.7 Simulated probabilities of errors in type recognition of 11 objects (curves 1 and 2) and of 6 objects (curves 3 and 4) versus bandwidth. The results are obtained for correlation processing with $\Gamma = 3$ individualized standard RPs (curves 1 and 3) and for cpdf processing (curves 2 and 4). The SNR was 25 dB; other parameters are the same as for Figure 4.6.

Target Class Recognition Using Individualized Standard RPs and cpdf of RPs. If the probability of error in recognition is too high, it can be decreased by combining various target types that are close to one another by some criterion into the same class, preserving, however, the individualized standards. Misrecognitions within close target types then become insignificant. Figure 4.8 shows the probability of error in recognition of four target classes: large-sized (Tu-16, B-52, B-1B, An-26 aircraft), medium-sized (Mig-21, F-15, and Tornado aircraft and an AH-64 helicopter), small-sized (missiles of ALCM and GLCM types), and a passive decoy target for single illumination

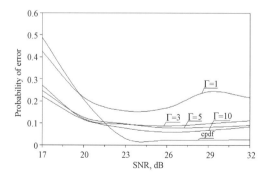

Figure 4.8 Simulated probabilities of error in recognition of four classes of the targets with individualized standard RPs: large-sized (Tu-16, B-52, B-1B, An-26), medium-sized (Mig-21, F-15, Tornado, AH64), and small-sized (ALCM, GLCM, and passive decoy) targets. Other parameters are the same as in Figure 4.6.

by a chirp signal of 80-MHz bandwidth. The target types of Figure 4.6 were combined here in classes by the criterion of similarity in radial extent. For the use of three specialized standard RPs on a type and SNR of 23 dB, the probability of error for single target illumination, equal to about 0.21 for type recognition, decreases to about 0.095 for the class recognition. Let us note that the An-26 aircraft is of such radial extent that including it in both classes of large- or medium-sized targets is legitimate. The curves of Figure 4.8 do not change notably if An-26 aircraft is included in the class of medium-sized targets.

4.1.9   Simulation of Target Type and Class Recognition Using Rotational Modulation of a Narrowband Signal

Let us proceed to the consideration of the sampled amplitude-frequency spectrum $G(F_k)$ (k = 1, 2, . . . , K) of a burst of pulses having a constant PRF and subjected to rotational modulation (Sections 4.1.5 and 4.1.6). The RMS can be subjected to correlation or cpdf processing, in particular, on the basis of the simplest and individualized standard RMSs and cpdf of RMSs.

The best results in this approach can be obtained for coherent radar with high PRF and long burst duration, but recognition will be simulated below for a coherent radar with moderate PRF and burst duration. It will be shown that some recognition information can be obtained in this case as well (provided that reliable engine parameters were substituted for our tentative ones).

Target Type and Class Recognition Using Individualized Standard RMSs and cpdf of RMSs. A narrowband signal at a wavelength λ = 5 cm consisting of 64 pulses with PRF of 1 kHz was simulated to produce the RMSs in the aspect sector of 20° to 40° that can provide a better recognition than the aspect sector of 0° to 20°.

Figure 4.9 shows the simulated probabilities of errors in target type recognition, with equiprobable appearance of nine objects, versus potential SNR in dB for various recognition algorithms:

- The correlation approximation of the additive Bayesian algorithm with Γ = 3 individualized standard RMSs in the aspect sector and with filtration-like centering (4.17);

- An additive Bayesian algorithm with direct evaluation of the cpdf of the RMSs. The target's doppler frequency was considered as exactly known in this case.

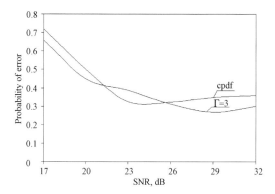

Figure 4.9 Simulated probabilities of errors in the RMS' recognition of nine objects for aspect sector 20° to 40° versus potential SNR for a single target illumination by narrowband signal (λ = 5 cm) consisting of 64 pulses with 1-kHz PRF. The Γ = 3 curve corresponds to correlation processing with three individualized standard RMSs on a type and with filtration-kind centering. The cpdf curve corresponds to cpdf processing and exactly known doppler frequency.

An estimate of the RMS cpdf was obtained using a Parzen window and the training sets of T = 300 RMSs for each target. The training was carried out in the presence of noise, depending on the potential SNR to be arbitrarily chosen for each RMS from 20 to 30 dB. In the processes of training and testing the target pitch angle was changed monotonically from −1° to 9° and the target roll angle was changed from −5° to 5°.

Each decision was made using the only RMS (N = 1) superimposed on the realizations of noise different from realizations used in the training. The simulated targets were the same as for Figure 4.6. But it was expedient to combine all the targets producing no evident rotational modulation, such as the missiles ALCM and GLCM and a passive decoy, into the class without engine (WE). The other eight types of targets (Tu-16, B-52, B-1B, An-26, Mig-21, F-15, Tornado aircraft, and AH-64 helicopter) are considered as the independent types. The case of a passive decoy with imitated rotational modulation was not considered yet. The results of simulation can be appreciably improved by uniting the target types, being close to one another, into the classes.

Target Class Recognition Using the Individualized Standard RMSs and cpdf of RMSs. We can introduce three classes of targets: the class of turbo-jet targets (Tu-16, B-52, B-1B, F-15, Tornado, and Mig-21), the class of turbo-prop and propeller targets (An-26 and AH-64), and the WE class. The probabilities of error in recognition are additionally decreased (Figure 4.10) at the expense

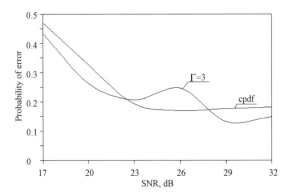

Figure 4.10 Simulated probabilities of errors in the recognition of three target classes versus potential SNR for a single (*N* = 1) target illumination by a narrowband signal (λ = 5 cm) consisting of 64 pulses with 1-kHz PRF. Other parameters are the same as in Figure 4.9.

of decreasing the recognition information measure, as will be shown below. Let us note here that these results must be considered critically, because there are dropouts in rotational modulation caused by the aspect change (Section 3.3). Hence, the target observed with such a dropout can be misrecognized as having no engine, and classes introduced here do not correspond to those introduced above.

4.1.10 Evaluation of Information Measures for Various Recognition Signatures and Their Combinations

Evaluation of information measures is desirable for comparing various signatures and alphabets of recognized objects (classes or types of targets). The method of such evaluation was considered already in Section 1.6. It allows the introduction of the concept of an alphabet of recognized objects to be optimal from an informational viewpoint. Indeed, to increase the conditional probabilities of the target recognition, one can combine them into classes. But this combining decreases the number of alternatives and the quantity of information. Therefore, such a decomposition on classes can be found that provides the maximum information quantity. The alphabets of recognized objects corresponding to such decomposition can be named the *optimal from informational viewpoint*. The alphabet, being optimal from informational viewpoint, need not be, but can sometimes be, the optimal from the viewpoint of a set of quality criteria. Our task is only to show the principal possibility of discussing such problems on the basis of backscattering simulation. We will limit ourselves in this discussion by the cases of single target illumination.

Figure 4.11 shows simulated information measures for the RP signature versus potential SNR:

- In recognition of 11 target types according to Figure 4.6;
- In recognition of four target classes according to Figure 4.8.

It is seen that the information measure is better for the type recognition (Figure 4.6). We noted above that when using three specialized standard RPs on a type and SNR of 23 dB, the probability of error diminishes from 0.21 in the case of type recognition to 0.095 in the case of class recognition. The information measure, then, decreases correspondingly from 2.2 to 1.45 bits. Therefore, from the informational viewpoint the alphabet of 11 target types is preferable to that of four target classes. One can note this and decide between the two alternatives.

But the type recognition is not always optimal even from the informational viewpoint. For example, the recognition of 11 target types by the

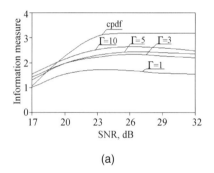

(a)

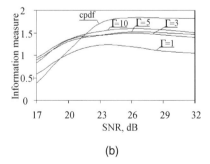

(b)

Figure 4.11 Simulated information measures of RP signature versus potential SNR in recognition of (a) 11 target types according to Figure 4.6, and (b) 4 target classes according to Figure 4.8.

RMS provides a probability of error of about 0.45 and an information measure of about 1.7 bits for the SNR of 23 dB and use of the cpdf algorithm. Creation of the WE class reduces the probability of error to about 0.32 and slightly increases the information measure to about 1.8 bits. If all the targets are included in four classes (turbo-jet, turbo-prop, propeller, and WE), then the probability of error is decreased further to about 0.2, and the information measure is reduced sharply to about 0.9 bits. Therefore, for the considered aspect sector of 20° to 40° the alphabet of nine recognized targets (Figure 4.9) is optimal from the informational viewpoint.

Another example of an alphabet optimal from the informational viewpoint can be given for recognition using only the trajectory signatures, the *V-H* ones. It includes six target classes. The first class includes medium-speed aircraft of Tu-16 and B-52 types and a medium-speed passive decoy; the second class includes the high-speed aircraft of B-1B, Mig-21, and Tornado types; the third class includes missiles of ALCM and GLCM types. F-15 and An-26 aircraft and the AH-64 helicopter are recognized as independent types. For such decomposition the information measure of recognition by the *V-H* signature constitutes 0.52 bits.

Comparing the optimal alphabets by their information measures for single target illumination, we can see them to be unequal for various signatures. Under adopted conditions we found that the information measure constituted 2.2 bits for the RP signature, 1.8 bits for the RMS signature, and 0.52 bits for the trajectory signature with the notes made above.

We did not mention here numerous cases, where on the one hand several signatures were used together, and on the other hand the contribution of some information measures decreased under some conditions. The aim of this material is to develop the informational approach to recognition that can be used in solving practical tasks.

4.2 Nonparametric Recognition Algorithms

Nonparametric algorithms were developed heuristically for unknown statistical distributions of recognition signatures. They include the algorithms of distance evaluation (Section 4.2.1), as well as voting algorithms (Section 4.2.2). After the review of these algorithms, examples of their simulation are presented (Section 4.2.3).

4.2.1 Recognition Algorithms of Distance Evaluation

The distance evaluation recognition algorithms include the algorithms of minimum distance, and of nearest neighbor and nearest neighbors.

Algorithms of Minimum Distance. They are developed for signature spaces $\boldsymbol{\alpha} = \|\alpha_\nu\|$, $\nu = 1, 2, \ldots N$ with some distance (dissimilarity) measure d_k. This measure can be defined as an interval between the point $\hat{\boldsymbol{\alpha}}_y$, corresponding to the estimate of parameter $\boldsymbol{\alpha}$ received from the current signal-plus-noise realization $\mathbf{y}$, and the point $\boldsymbol{\alpha}_k$, corresponding to the mean value or directly to one of the training values of parameter $\boldsymbol{\alpha}$ for the actual presence of the kth object ($k = 1, 2, \ldots, M$).

It is supposed that all the necessary values $\boldsymbol{\alpha}_k$ have been stored in a computer memory at the training stage. Two equivalent forms of the minimum distance algorithms are usually used

$$i = \hat{k}_{\text{opt}} = \arg\min_k d_k \qquad (4.20)$$

$$i = \hat{k}_{\text{opt}} = \arg\min_k d_k^2 \qquad (4.21)$$

There are basic variants of squared distance measures [2]:

- For uncorrelated signatures with equal variances

$$d_k^2 = \left| \hat{\boldsymbol{\alpha}}_y - \boldsymbol{\alpha}_k \right|^2 = (\hat{\boldsymbol{\alpha}}_y - \boldsymbol{\alpha}_k)^{\text{T}}(\hat{\boldsymbol{\alpha}}_y - \boldsymbol{\alpha}_k) = \sum_{\nu=1}^{N} (\hat{\alpha}_{y\nu} - \alpha_{k\nu})^2$$
$$(4.22)$$

- For correlated signatures with arbitrary variances

$$d_k^2 = (\hat{\boldsymbol{\alpha}}_y - \boldsymbol{\alpha}_k)^{\text{T}}\boldsymbol{\Phi}^{-1}(\hat{\boldsymbol{\alpha}}_y - \boldsymbol{\alpha}_k) \qquad (4.23)$$

where $\boldsymbol{\Phi}$ is the correlation matrix of signatures.

Diagonalization $\boldsymbol{\Phi} = \mathbf{U}\boldsymbol{\Lambda}\mathbf{U}^{-1}$ of this matrix is frequently used, where $\boldsymbol{\Lambda}$ is the diagonal matrix of eigenvalues and $\mathbf{U}$ is the unitary matrix. The use of orthogonal generalized signatures $\boldsymbol{\xi} = \boldsymbol{\Lambda}^{-1/2}\mathbf{U}^{-1}\boldsymbol{\alpha}$ is also possible. For this case (4.23) becomes the variant of (4.22):

$$d_k^2 = (\hat{\boldsymbol{\xi}}_y - \boldsymbol{\xi}_k)^{\text{T}}(\hat{\boldsymbol{\xi}}_y - \boldsymbol{\xi}_k) \qquad (4.24)$$

The overall number of signatures can be reduced in this case simply by rejecting the part that has small eigenvalues [3].

Algorithms of Nearest Neighbors [4]. They envisage the possibility of introducing into a computer memory the training estimates $\gamma = 1, 2, \ldots, T_k$ of a vector parameter $\hat{\boldsymbol{\alpha}}_{\gamma|k}$, which are used here as the signatures of training statistics assigned for recognition of various target classes or types k. After evaluating the estimate $\hat{\boldsymbol{\alpha}}_\gamma$ from the current "signal-plus-noise" realization **y**, the L nearest to its neighbors $\hat{\boldsymbol{\alpha}}_{\gamma|k}$ are found, which belong, generally speaking, to objects of various classes (types). The observed object is recognized as belonging to the same ith class (type) as that to which most of its L nearest neighbors $\boldsymbol{\alpha}_{\gamma|k}$ belong. The degree of nearness can be evaluated using one of the distance measures (4.22) through (4.24). If the number $L = 1$, then the algorithm of the nearest neighbors is reduced to the algorithm of the nearest neighbor.

4.2.2 Recognition Voting Algorithms

Recognition voting algorithms include the algorithms of weighted voting, of simple voting, and of multiple voting.

The Algorithm of Weighted Voting [10]. It combines the preliminary decisions i_ν received on the basis of different signatures $\nu = 1, 2, \ldots, N$, taking into account the degree of their reliability. The structure of the weighted voting algorithm

$$i = \hat{k} = \arg \max_k \left[\sum_{\nu=1}^{N} P_k(i_\nu) + \ln P_k \right] \qquad (4.25)$$

is analogous to the additive Bayesian algorithm (4.12), but it is comparably simpler. Discrete distributions $P_k(i_\nu)$ accounting for reliability of preliminary decision numbers i_ν are used here instead of continuous distributions $p_k(\hat{\boldsymbol{\alpha}}_\nu)$ of parameter estimates $\hat{\boldsymbol{\alpha}}_\nu$ included in expression (4.12). Exclusion of values r_k from (4.25) corresponds to the simple cost matrix used in Sections 1.6.1 and 4.1.1. The matrices $\mathbf{P}(\nu) = \left\| P_k(i_\nu) \right\| = \left\| P_{ki}(\nu) \right\|$ of $K \times K$ dimension [where K is the number of the target classes (types)] are supposed to be known from an experiment or simulation for all the signatures ν and stored in a computer memory. After each preliminary decision i_ν has been obtained using the νth signature, the ith column of the corresponding νth matrix $\mathbf{P}(\nu)$ is extracted from the computer memory so as to compare the expressions in square brackets of (4.25) for various hypotheses k.

The Simple Voting Algorithm [10]. It combines the decisions made on the basis of different signatures without detailed consideration of their reliability.

Such simplification is justifiable only for approximately equiprobable errors of the decisions inherent to the combined signatures [compare with the case of the simplest distance measure (4.22)]. In this case the identity matrices $\mathbf{I}(\nu) = \left\| \delta_{ki}(\nu) \right\|$, where $\delta_{ki}(\nu) = 1$ for $i = k$ and $\delta_{ki}(\nu) = 0$ for $i \neq k$, replace in (4.24) the matrices $\mathbf{P}(\nu) = \left\| \ln P_k(i_\nu) \right\| = \left\| P_{ki}(\nu) \right\|$ of general type. The possible inequality of a priori values $\ln P_k$ for various hypotheses k are also neglected, so that

$$i = \hat{k} = \arg \max_k \sum_{\nu=1}^{N} \delta_{ki}(\nu) \qquad (4.26)$$

The Multiple Voting Algorithm [11]. A large data set is decomposed onto smaller subsets. If these subsets are too large, they can be decomposed onto still smaller subsets. The simple voting within each data subset, or primary voting, is provided at the initial stage of decision-making. The next stage, secondary voting, uses the results of the primary voting, and, if necessary, tertiary voting is also carried out.

4.2.3 Simulation of Nonparametric Recognition Algorithms

Simulation of Algorithms of Voting, Minimum Distance with the Simplest Distance Measure, and Additive Bayesian [9]. As in Section 4.1.7, radar recognition of four targets' classes [(1) large-sized aircraft, (2) medium-sized aircraft, (3) missile, and (4) passive decoy missile] was considered. The set of signatures SS1 (Section 4.1.7) included the maximum of correlation sums of the single ($N = 1$) input RP calculated for $\Gamma = 3$ simplest standard profiles on each class, and wideband RCS. The standard RPs were obtained by a chirp signal of 80-MHz bandwidth in the high-frequency part of S-band for the aspect sector of $20°$ on the basis of several tens $T > \Gamma$ of simulated profiles (Section 2.2). Together with fluctuations of simulated RCS, an a priori error of its measurement (due to instabilities of transmitter power and receiver sensibility) was taken into account with a variance of 3 dB. The factor of imprecisely known range of a target was considered here also. Four algorithms of recognition by a set of signatures were simulated: (1) additive Bayesian algorithm (4.12) with correlation processing, (2) weighted voting algorithm (4.25), (3) algorithm of minimum distance with the simplest distance measure (4.20) and (4.21), and (4) simple voting algorithm (4.26). Figure 4.12 shows the simulated probabilities of errors in target class recognition, with equiprobable appearance of objects, versus potential SNR in dB for these recognition algorithms under the condition of a single ($N = 1$) target illumination.

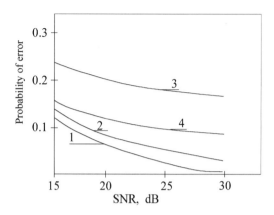

Figure 4.12 Simulated probabilities of errors in the target class recognition versus potential SNR for the data consolidation algorithms: (1) additive Bayesian algorithm,
(2) weighted voting, (3) minimum distance with the simplest distance measure,
and (4) simple voting. The single (N = 1) target illumination by an 80-MHz
chirp signal, correlation processing with Γ = 3 simplest standard RPs on a
class, and the use of wideband RCS as a signature are simulated [9].

Despite the simplification, the algorithm of weighted voting (4.25)
shows a performance near to that of the additive Bayesian algorithm (4.12)
with correlation processing (4.18). The performance of the algorithm of
minimum distance with the simplest distance measure (4.21) and (4.22)
and of the simple voting algorithm (4.26) were worse. The latter is due to
the equalized significance of the coarse RCS information and of the more
precise one related to the extent and shape of RPs used in the algorithms
(4.21) and (4.22), and (4.26).

*Comparative Simulation of the Nearest Neighbor Algorithm and Some Other
Algorithms [12].* By means of scaled electrodynamic simulation, a great set
of backscattering experiments was carried out to realize the recognition of
five models of aerial targets using coherent and noncoherent four-frequency
target illumination. The nearest neighbor algorithm, an algorithm identical
to the Bayesian cpdf one, and a neural recognition (see Section 4.4) algorithm
were used. All these algorithms provided close recognition quality for the
identical (coherent, noncoherent) signals. The nearest neighbor algorithm was
not behind other algorithms. We must notice only that the real instabilities of
amplitude information due to unaccounted-for instabilities of propagation
conditions and hardware instabilities are not usually considered when using
the nearest neighbor algorithm.

Use of the Simple Voting Algorithms in the Closing Stage of RP Processing [13]. A data set of 11,968 normalized RPs belonging to aerial targets of 24 types was obtained from a real 300-MHz bandwidth radar in S-band for nine aspect sectors of 20°. The set obtained for each sector was then divided into groups of bursts consisting of 12 bursts with a scan period of 2 seconds. Each burst consisted of eight RPs. Bursts were numbered within each group, and the odd bursts were then related to the training set, and the even bursts to the test set. Each "training" part of the group was used to form one standard RP and to build one corresponding matched filter (4.17), the total number of such filters amounting to 119 (including 33 filters for nose-on aspects). Each RP from the test part of the group was addressed to all the matched filters corresponding to the known aspect sector. In this initial stage (i.e., the stage of the correlation processing) about 1600 test RPs of nose-on aspect sector, representing 13 distinct aircraft types, were recognized, each with averaged error probability of 0.21.

In the closing stage of processing, two variants of processing were tested. Voting within only the burst decreased the error probability to 0.16 for nose-on aspects. Voting within the test part of the group (about 50 RPs) decreased the probability of errors to zero not only for the nose-on aspects (0° to 20°), but also for the aspects of 20° to 40°, 40° to 60°, 60° to 80°, 80° to 100°, and 140° to 160°. At several aspects the probability of errors remained appreciable and constituted 0.2 at the aspects of 100° to 120°, and 120° to 140°, and 0.05 at the aspects of 160° to 180°.

The work [13] published by Hudson and Psaltis was the first experimental work demonstrating the possibilities of recognizing not only classes but types of targets using RPs. However, the simple voting algorithm used in the closing stage of the RP processing does not belong to the optimal algorithms. Certainly, simple voting does not affect the recognition quality so much, as in Figure 4.12, because of equiprobable errors of partial decisions. But the number N of such decisions used is too great in this case to increase the quality of recognition. The results of simulation (Figure 4.6) suggest to us that the number of standards 33/(24 to 13) = 1 to 2 is apparently too small in this case. Some interval introduced between the RPs to provide their independence could be also justified.

4.3 Recognition Algorithms Based on the Precursory Data Transform

Definite attention was recently paid to the initial data transform precursory to its storage, transmitting, processing [14], and recognition, for instance

[15]. A special issue of the *IEEE Proceedings* [14] was devoted to the use of the wavelet transform. The wavelet transform has been included into application program packages such as "Mathcad." Certain information about wavelets, therefore, is included in this book also (Section 4.3.1). The discrete wavelet transform and its use in recognition are considered in Section 4.3.2. Simulation examples of recognition algorithms with the wavelet transform are presented in Section 4.3.3.

4.3.1 Wavelet Transform and Wavelets

The wavelet transform is a kind of generalized Fourier transform. Decompositions of an arbitrary function onto orthogonal harmonics and onto the orthogonal sinc-functions are special cases of such transforms, ensuring either high frequency (poor time) or high time (poor frequency) resolution. The wavelet transform is designed to provide comparatively good common time-frequency resolution [14, 16].

Wavelet functions (wavelets) used in the wavelet transform are chosen as

$$\psi_{k,q}(t) = 2^{k/2}\,\psi[2^k t - q] \quad (k = 0, 1, \ldots K - 1; \; q = 0, 1, \ldots 2^{K-1} - 1)$$
$$(4.27)$$

where a function $\psi(t)$ is known as a "mother" wavelet. The parameter k introduces time scale compression of the mother wavelet by $2^0, 2^1, \ldots, 2^{K-1}$ times. The parameter q introduces time shifts of the compressed wavelets. The multiplier $2^{k/2}$ normalizes the square of wavelet $\psi_{k,q}(t)$ to unity in the given interval of the argument t if the square of the mother wavelet was already so normalized in this interval.

Ingrid Daubechies wavelets are a kind of wavelet-function $\psi_m(t)(m = 2^{k-1} + q)$, built according to (4.27) beginning from $m = 2$. They are strictly limited in the time domain and not strictly limited in the frequency domain. Being orthonormal, they are useful in generalized Fourier transforms. Only digital Daubechies wavelets were introduced into the "Mathcad" program package. The Daubechies wavelets in Figure 4.13 for $m = 0$ (solid line) and $m = 1$ (dashed line) are not described by (4.27) unlike such wavelets for $m \geq 2$. Wavelets for $m = 17$, $m = 18$, and $m = 20$ ($q = 0$, $q = 1$, and $q = 4$ for $k = 4$) produced on the basis of a "mother" wavelet $m = 2$ are presented in Figure 4.14. For an arbitrary value of K, the digital Daubechies wavelets form the orthogonal matrix ψ of $2^K \times 2^K$ dimension.

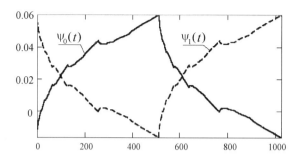

Figure 4.13 Daubechies wavelets for *m* = 0 (solid line) and *m* = 1 (dashed line).

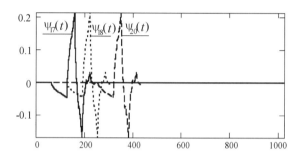

Figure 4.14 Daubechies wavelets for *m* = 17, *m* = 18, and *m* = 20.

4.3.2 Discrete Wavelet Transform and Its Use in Recognition

The direct wavelet transform of the $M = 2^K$-dimensional vector $\mathbf{y} = \left\| y_0 \quad y_1 \quad y_{M-1} \right\|^T$ is of the matrix form

$$\mathbf{g} = \left\| g_0 \quad g_1 \quad \cdots \quad g_{M-1} \right\|^T = \boldsymbol{\psi}\mathbf{y} \qquad (4.28)$$

analogous to that of the discrete Fourier transform. Operation (4.28) can be fulfilled in "Mathcad" by its **wave**(**y**) function.

The inverse discrete wavelet transform of the $M = 2^K$-dimensional vector

$$\mathbf{g} = \left\| g_0, g_1, \cdots g_{M-1} \right\|^T$$

is of the matrix form, analogous to that of the discrete inverse Fourier transform,

$$y = \boldsymbol{\psi}^{-1}\mathbf{g} = \boldsymbol{\psi}^{T}\mathbf{g} \qquad (4.29)$$

The property of orthogonality of the matrix

$$\boldsymbol{\psi}^{-1} = \boldsymbol{\psi}^{T} \qquad (4.30)$$

was used in (4.29). Operation (4.29) can be fulfilled in "Mathcad" by use of its *iwave* (**g**) function.

It is interesting that vector-functions *iwave* ($\| 1 \quad 0 \quad \ldots \quad 0 \|^{T}$), *iwave* ($\| 0 \quad 1 \quad 0 \quad \ldots \quad 0 \|^{T}$), and *iwave* ($\| 0 \quad 0 \quad \ldots \quad 0 \quad 1 \|^{T}$) define the first, second, . . . , and the last sampled wavelet.

Correlation Sums on the Wavelets' Basis. The correlation sum of (4.15) type used in recognition can be transformed by use of the presented expressions (4.28) through (4.30):

$$Z_{\Sigma} = \mathbf{Y}^{T}\mathbf{X} = \mathbf{Y}^{T}\boldsymbol{\psi}^{-1}\boldsymbol{\psi}\mathbf{X} = (\boldsymbol{\psi}\mathbf{Y})^{T}(\boldsymbol{\psi}\mathbf{X}) = \mathbf{g}_{y}^{T}\mathbf{g}_{x} \qquad (4.31)$$

The latter means that the correlation sum can be calculated after the transformation into wavelet domain [16].

4.3.3 Simulation of Wavelet Transforms and Evaluation of Their Applicability in Recognition

Simulation of the RPs' Wavelet Transforms. Figure 4.15(a) shows simulated RPs, and Figure 4.15(b) shows corresponding wavelet-RPs (WRPs) of the Tu-16 and Mig-21 aircraft and ALCM missile for chirp illumination of 80-MHz bandwidth. Transformation of vectors **y** corresponding to the RPs into vectors **g** corresponding to the wavelet-RPs is provided according to (4.28) and practically realized by use of the *wave* (**y**) function of the "Mathcad" program.

In Table 4.2 the simulated maximum correlation factors $\rho_{\mu,\nu}$ of WRPs μ, $\nu = 1, 2, 3$ [Figure 4.15(a)] are shown for a (1) large-sized target, (2) medium-sized target, and (3) small-sized target. Analogous data for the RPs are presented in Table 4.3. Maximum in both tables is understood with regard to (4.17). It can be seen that both kinds of calculating the correlation sums give the same results according to (4.31).

It can also be seen from Tables 4.2 and 4.3 that differences in the correlation factor of different targets are not large. It clarifies the necessity

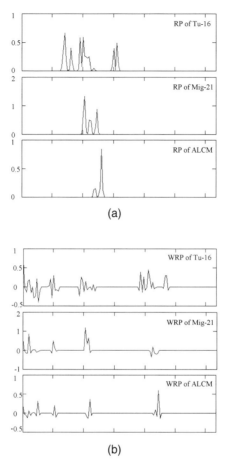

(a)

(b)

Figure 4.15 Simulated (a) RPs and (b) WRPs of the Tu-16 and Mig-21 aircraft and a ALCM missile for 80-MHz bandwidth chirp illumination.

Table 4.2
Correlation Coefficients for WRPs

ν＼μ	1	2	3
1	1	0.485	0.495
2	0.485	1	0.735
3	0.495	0.735	1

Table 4.3
Correlation Coefficients for RPs

μ ν	1	2	3
1	1	0.485	0.495
2	0.485	1	0.735
3	0.495	0.735	1

to use noticeably greater SNR in recognition than in detection, as it was calculated above in Sections 4.1 and 4.2.

In Figure 4.16 the WRPs of Tu-16 aircraft are shown for close but different ranges. It can be seen that the WRP structure changes with the shift of RP's position within the range gate, thus complicating the WRP's centering.

About the Applicability of Wavelet Transform in Recognition. This transform does not degrade the information; therefore, it can be used in recognition. Under our limited simulation, however, we found no advantages in using the wavelet transform for RP processing.

4.4 Neural Recognition Algorithms

Neural recognition algorithms have broad enough possibilities and attract serious attention due to the simplicity of initial data presentation. We review

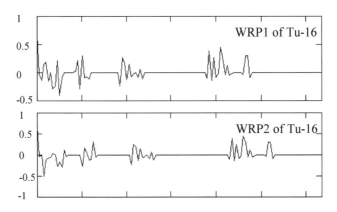

Figure 4.16 Simulated WRPs of the Tu-16 aircraft for close but different ranges.

the structures and optimization criterion for artificial neural networks (ANNs) in Section 4.4.1. The gradient algorithms for ANNs training are given in Section 4.4.2. In Sections 4.4.3 through 4.4.5 we present the results of the gradient algorithms operation, including the specifics of target class and type recognition. The evolutionary (genetic) algorithms for ANNs training, which have recently appeared, are reviewed in Section 4.4.6.

4.4.1 Structures and Optimization Criterion for ANNs

ANNs belong to the class of systems with a given structure and conditional optimization. This means that optimization is provided in correspondence to the chosen structure. Structures of ANNs are accepted on the analogy with biological neural nets, consisting of great sets of neurons with massive cross-connections. The ANN can have nonmodularized (ANN NM) and modularized (ANN M) substructures [17–24].

An artificial neuron (AN), as a node of ANN graph, performs nonlinear and linear operations

$$z = f(w + \beta), \quad w = \sum_s \alpha_s y_s \qquad (4.32)$$

Here, y_s, is the signal at its sth input, α_s is the connection weight of sth input, and β is the bias of this AN. Nonlinear operations are usually provided by one of the so-called sigmoid functions:

$$f_1(w) = \frac{e^w - e^{-w}}{e^w + e^{-w}}, \quad \text{or} \quad f_2(w) = \frac{1}{1 + e^{-w}}$$

These functions vary from -1 to 1 (the first) and from 0 to 1 (the second), when their argument w is changed from $-\infty$ to ∞. These functions are sometimes called activation functions and they operate as smooth thresholds. Thus, the output of each AN is activated gradually when the weighted sum of input signals w rises above this smooth threshold β. For large absolute values of w, these functions correspond to a hard limitation; for the smaller absolute values of w, the limitation is mild (smooth).

A feedforward artificial neural network, nonmodularized (FANN NM) consists of several layers of identical neurons, so that input data only flow forward from layer to layer. The widely used structure of ANN NM (Figure 4.17) consists of three neuron layers, two of which are the processing layers.

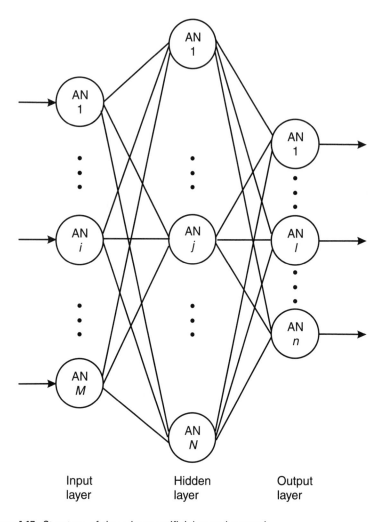

Figure 4.17 Structure of three-layer artificial neural network.

Representation of structure (Figure 4.17) by a single layer is also possible (Figure 4.18). As the study has shown [19], the structure of three layers is capable of solving various problems with a quality no worse than that of structures containing a larger number of layers.

Let us consider in more detail the three-layer ANN. The first layer of Figure 4.17 just transfers the input data onto the next layers without processing. Number M of elements in the first layer corresponds to the number of the scalar signatures used for recognition. Two other layers, the N-element

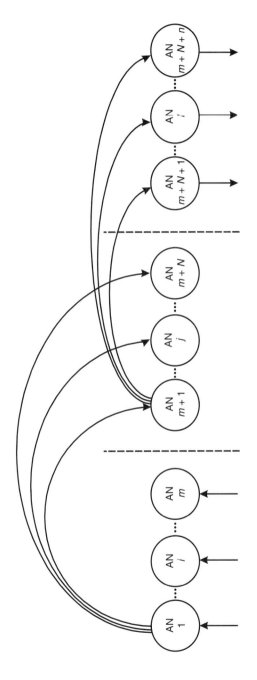

Figure 4.18 Representation of three-layer artificial neural network by a single layer.

and n-element ones, process and transfer the incoming data y_1, y_2, $\ldots$, y_M according to (4.32). So,

$$z_i = y_i, \qquad\qquad (1 \leq i \leq M) \qquad (4.33)$$

$$z_j = f(w_j + \beta_j), \qquad w_j = \sum_{i=1}^{M} \alpha_{ji} z_i, \qquad (1 \leq j \leq N) \qquad (4.34)$$

$$z_l = f(w_l + \beta_l), \qquad w_l = \sum_{j=1}^{N} \alpha_{lj} z_j, \qquad (1 \leq l \leq n) \qquad (4.35)$$

The number of output elements n is equal to the number of required recognition decisions. Adjustment of the connection weights is carried out so that the output value $z_{\hat{l}}(1 \leq \hat{l} \leq n)$, determining the desirable decision $\hat{l}$, approaches the level of about 0.9, and the other values z_l approaches the level of about 0.1. These levels are usually chosen instead of unity and zero levels so as to avoid very small derivatives of the sigmoid functions. The number N of intermediate (hidden) layer elements is usually appreciably greater than the sum $M + n$.

The FANN NM optimization criterion consists of minimization of the quadratic function of mean cost, with the multiplier of 1/2 introduced to simplify the subsequent relations:

$$r(\boldsymbol{\alpha}) = \sum_{p=1}^{P} \sum_{l=1}^{n} (z_{pl} - z_{l|p})^2 / 2 \qquad (4.36)$$

The cost function (4.36) evaluates the degree to which the actual outputs z_{pl} of FANN fit the desirable ones $z_{l|p}$ $(1 \leq l \leq n)$ over the training set of signatures y_{p1}, y_{p2}, $\ldots$, y_{pi}, $\ldots$, y_{pM} $(1 \leq i \leq M)$ which form the M-element vectors of signatures $\mathbf{y}_p$ $(1 \leq p \leq P)$. The training set must include various combinations of signatures for all the objects to be recognized by the FANN. The desirable outputs $z_{l|p}$ are the preset values to be achieved after presentation of each pth training realization for the object of a known type. This minimization has to achieve an optimal value of the parameter's vector $\boldsymbol{\alpha}$, containing all the connection weights α_{ji} and biases β_j of the hidden layer (4.34), and those α_{lj} and β_l of the output layer (4.35).

Modularity in neural computing indicates a possible specialization of the ANN's parts (modules), according to their functional destination, and is a subject of modern investigations. The modular approach is reasoned by analogies with the cerebral cortex structure of vertebrates [20]. Modularity corresponds also to the structure of the Bayesian algorithm (4.6) through (4.12). But excessive modularity leads to sophisticated processing.

Simulation of recognition can show an advisability and rational degree of modularity in each given case.

4.4.2 Gradient Algorithms for Training the FANN

The connection weight training of the FANN consists of parameter optimization in the course of presentation of lots of the objects' instances. Simple gradient, pair gradient [17], evolutionary (genetic) algorithms [23] (Section 4.5.3), and some others [17] are used to train various FANN.

Simple Gradient Algorithm (SGA) for FANN NM Training. Components of the parameter vector $\boldsymbol{\alpha}_k = \left\| \alpha_{k1} \quad \alpha_{k2} \quad \ldots \quad \alpha_{ks} \quad \ldots \quad \beta_{ks} \quad \ldots \right\|^{\mathrm{T}}$ at the initial iteration $k = 0$ can be chosen arbitrarily. Here, s is a current number of parameter without regard to the layer in which it is placed. Then the parameter vector $\boldsymbol{\alpha}_k$, $k = 0, 1, 2, \ldots$ obtains successive increments

$$\Delta_k \boldsymbol{\alpha}_k = \boldsymbol{\alpha}_{k+1} - \boldsymbol{\alpha}_k = -\gamma \, dr(\boldsymbol{\alpha}_k)/d\boldsymbol{\alpha}, \quad (k = 0, 1, 2, \ldots) \quad (4.37)$$

where $dr(\boldsymbol{\alpha})/d\boldsymbol{\alpha}$ is the gradient of function $r(\boldsymbol{\alpha})$, and γ is a coefficient, small usually, that is selected in the training process.

The iteration process can be reconciled with the presentation of instances of various class objects $p = 1, 2, \ldots, P$. Then increments of the parameter vector for each object's instance can be evaluated analogously

$$\Delta_p \boldsymbol{\alpha}_p = -\gamma \, dr_p(\boldsymbol{\alpha}_p)/d\boldsymbol{\alpha}, \quad (p = 0, 1, 2, \ldots, P) \quad (4.38)$$

where

$$\boldsymbol{\alpha}_p = \left\| \alpha_{p1} \quad \alpha_{p2} \quad \ldots \quad \alpha_{ps} \quad \ldots \quad \beta_{ps} \quad \ldots \right\|^{\mathrm{T}},$$

and

$$r_p(\boldsymbol{\alpha}) = \sum_{l=1}^{n} (z_{pl} - z_{l|p})^2 / 2 \quad (4.39)$$

Increments of each component α_{ps} and β_{ps} of the parameter vector $\boldsymbol{\alpha}$ can also be found from (4.38) as follows:

$$\Delta_p \alpha_{ps} = -\gamma \, dr(\boldsymbol{\alpha}_p)/d\alpha_{ps}, \quad \Delta_p \beta_{ps} = -\gamma \, dr(\boldsymbol{\alpha}_p)/d\beta_{ps} \qquad (4.40)$$

Presenting the object instances from the training set $p = 0, 1, 2, \ldots$, P, reiterating their presentation and using (4.39) to adjust the connection weights and biases, one can recursively evaluate the proper values of the parameter vector $\boldsymbol{\alpha}$.

Evaluation of SGA Parameters in FANN NM Training. Using (4.35), (4.39), (4.40) and differentiation rules, one can find increments of the output layer parameters ($1 \le l \le n$):

$$\Delta_p \alpha_{lj} = -\gamma \frac{\partial r_p}{\partial z_l} \frac{\partial z_l}{\partial w_l} \frac{\partial w_l}{\partial \alpha_{lj}} = \gamma \delta_{pl} z_j, \quad \Delta_p \beta_l = -\gamma \frac{\partial r_p}{\partial z_l} \frac{\partial z_l}{\partial \beta_l} = \gamma \delta_{pl}$$
$$(4.41)$$

where

$$\delta_{pl} = f'(w_l + \beta_l)(z_{l|p} - z_{pl}) \qquad (4.42)$$

Using (4.34), (4.35), (4.39), (4.40), (4.42), and the differentiation rules, one can find increments of the hidden layer parameters ($1 \le j \le N$):

$$\Delta_p \alpha_{ji} = -\gamma \sum_{l=1}^{n} \frac{\partial r_p}{\partial z_l} \frac{\partial z_l}{\partial w_l} \frac{\partial w_l}{\partial z_j} \frac{\partial z_j}{\partial w_j} \frac{\partial w_j}{\partial \alpha_{ji}} = \gamma d_{pj} y_{pi} \qquad (4.43)$$

$$\Delta_p \beta_{ji} = -\gamma \sum_{l=1}^{n} \frac{\partial r_p}{\partial z_l} \frac{\partial z_l}{\partial w_l} \frac{\partial w_l}{\partial z_j} \frac{\partial z_j}{\partial \beta_j} = \gamma d_{pj} \qquad (4.44)$$

where

$$d_{pj} = f'(w_j + \beta_j) \sum_{l=1}^{n} \delta_{pl} \alpha_{lj} \qquad (4.45)$$

The derivative of function $z = f(w) = (1 + e^{-w})^{-1}$ can be used in the form $f'(w) = z(1 - z)$.

The Pair Gradient Algorithm (PGA) for FANN NM Training. This is designed to develop the simple gradient algorithm (4.38) by accounting for two successive gradients with various weights γ and η, so that

$$\Delta_p \boldsymbol{\alpha}_p = -\gamma \, dr_p(\boldsymbol{\alpha}_p)/d\boldsymbol{\alpha} - \eta \, dr_{p-1}(\boldsymbol{\alpha}_{p-1})/d\boldsymbol{\alpha}, \quad (p = 0, 1, 2, \ldots, P) \tag{4.46}$$

Increments (4.41) for the parameters of output layer then become

$$\Delta_p \alpha_{lj} = \gamma \delta_{pl} z_{pj} + \eta \delta_{(p-1)l} z_{(p-1)j}, \quad \Delta_p \beta_j = \gamma \delta_{pl} + \eta \delta_{(p-1)l} \tag{4.47}$$

and equations (4.43), (4.44) for the parameters of intermediate layer become

$$\Delta_p \alpha_{ji} = \gamma d_{pj} y_{pi} + \eta d_{(p-1)j} y_{(p-1)i}, \quad \Delta_p \beta_j = \gamma d_{pj} + \eta d_{(p-1)j} \tag{4.48}$$

where δ_{pl} and d_{pj} are defined by (4.42) and (4.45).

4.4.3 Simulation of Target Class Recognition Using Neural Algorithm with Gradient Training

Conditions of Simulation. In most cases of simulation the recognized classes were the large-sized (Tu-16, B-52, B-1B) and medium-sized (Mig-21, F-15, Tornado) aircraft and cruise missiles (ALCM, GLCM). In the last case of simulation, we considered passive decoy as a fourth class. With the purpose to investigate the ANN's potential to provide robust recognition, the following factors were accounted for in simulation [21]:

- Uncertainty in potential signal-to-noise ratio (SNR);
- Starting (sector 0°–10°) and increased (sector 0°–50°) aspect uncertainty;
- Range uncertainty (of several range resolution intervals).

We mainly used the RP as the signature. To recognize four target classes including passive decoy, we used the combination of the RP with wideband RCS. The passive decoy was assumed to have a radial dimension close to that of a missile, so that its normalized RPs did not differ from those of the missiles (more than one reflector can be installed on the decoy). Information about the RCS was introduced into the RP by its corresponding scaling.

The RPs were simulated counting on the chirp signal with a bandwidth of about 80 MHz. These RPs were observed in a range gate some wider than 64m with the samples taken 1m apart (the samples were two times more frequent than those corresponding to the boundary values of the sampling theorem). The number of input nodes of the ANN was $M = 64$. To account for range uncertainty, the centering was provided. The median sample Y_{med} of each RP was found for this purpose and aligned with the center of the input layer limited by M nodes. The number of output nodes n was equal to the number of the target classes $K = 3$ or $K = 4$. The number of hidden nodes varied within limits of 80 to 200 in accordance with the factors accounted for (listed above). Probabilities of error in recognition of target classes are estimated below counting on a single target illumination.

Class Recognition by Means of RPs (the Case of Range, SNR, and Aspect Uncertainty). As it was noted above, the range uncertainty here and below was accounted for by RP centering. The aspect uncertainty was accounted for by selecting the training RPs from the 0° to 10° aspect sector. To account for the SNR uncertainty, an automatic gain control (AGC) was simulated providing constant false alarm rate (CFAR) with the noise variation set to unity. The training was carried out in the presence of noise for several SNRs of 20, 22, and 24 dB using the sets of 60 RPs for each of $K = 3$ classes. The number of hidden layer nodes was 80.

The recognition performance for each of the three classes was tested using the sets of 1600 RPs on a class as the SNR changed from 18 to 30 dB. Probability of error in class recognition P_{er} for this complex case was 0.06.

Class Recognition by Means of RPs (the Case of Range and Increased Aspect Uncertainty). A single ANN for aspect sector of 0° to 50° was used instead of five ones of 10°. The number of the target 10° aspect sector was in this case estimated and applied to one of ANN's inputs together with its RP, both at training and testing stages. Training was carried out for $K = 3$ classes with SNR of 20 dB. The number of hidden layer nodes was 120. The ANN structure for this case is shown in Figure 4.19.

The recognition performance was tested using the sets of 300 RPs for each of aircraft classes and of 200 RPs for missile class in each of five 10° aspect sectors. Probability of error in class recognition P_{er} for this complex case was 0.08.

Class Recognition by Means of RPs (the Case of Range, SNR, and Increased Aspect Uncertainty). The possibility of simultaneous accounting for the SNR (in the interval of 18 to 30 dB) and increased aspect uncertainty (in the sector of 0° to 40°) was also investigated for the case of $K = 3$ target class

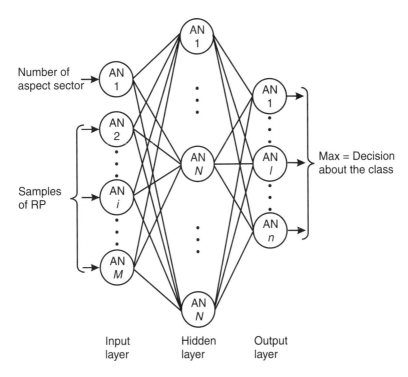

Figure 4.19 Variant of ANN structure for target recognition in the case of increased aspect uncertainty.

recognition. Training was carried out for three fixed SNRs of 20, 22, and 24 dB in four aspect sectors; the numbers were applied to one of ANN's inputs (as in Figure 4.19). The training set was then enlarged to 2160 RPs; the number of hidden nodes was also increased to 180 relative to the previous example.

The recognition performance was tested for aspect sector of 0° to 40° using 16,000 RPs and designating the SNRs from interval of 18 to 30 dB. Probability of error in class recognition in this most complex case increased in comparison with the previous case and became equal to 0.09.

Class Recognition by Means of RPs and Wideband RCSs (the Case of Range, SNR, and Aspect Uncertainty). For recognition of $K = 4$ classes, the RCS information was assumed to be simulated with artificially introduced instabilities of 3 dB. Training of the ANN was carried out in aspect sector of 0°–10° for the SNRs of 18, 21, 24, and 27 dB. The overall number of training RPs of four classes was 960. The numbers of hidden and output nodes were 200 and 4, respectively.

The recognition performance was tested using 5000 RPs. Probability of error in class recognition using a single RP was 0.04 for the SNRs of 18 to 30 dB. Despite recognition of an additional class, the probability of error is lower than in the previous case. We suppose that this is due to the use of additional information and the decrease of aspect uncertainty.

4.4.4 Simulation of Target Type Recognition Using Neural Algorithm with Gradient Training

Conditions of Simulation. The recognized targets were the Tu-16-, B-52-, B-1B-, Mig-21-, F-15-, Tornado-, and An-26-type aircrafts, ALCM- and GLCM-type missiles, the AH-64 helicopter, and a passive decoy. The SNR in simulation was assumed to be constant and equal to 25 dB. The aspect sector of targets was $0°$ to $10°$ from the nose.

We used as signatures the normalized RPs, so the information about RCS was not used. The RPs were obtained using a signal with bandwidth of about 80 MHz. These RPs were observed in a range gate some wider than 64m with samples taken 2m apart (samples corresponding to the boundary values of the sampling theorem). The positions of RPs in the range gate were assumed to be exactly known. This corresponds to high-quality target tracking using a wideband signal. The number of input nodes of the artificial neural network was $M = 32$. The number of output nodes n was equal to the number of the target types $K = 11$. The number of hidden nodes was 200. Probabilities of error in recognition of target types are estimated below counting on a single target illumination. The number of RPs for the ANN training was 990 (90 RPs for each target type). In the process of training and testing, the target pitch angle was changed monotonically from $-2°$ to $2°$ and the target roll angle was changed from $-3°$ to $3°$.

Type Recognition by Means of RPs (the Case of Given SNR and Aspect Sector). Figure 4.20 shows the probability of error in recognition of $K = 11$ target types versus SNR in dB. It can be seen that this probability is close to that obtained for the case of target type recognition using the Bayesian algorithm with the cpdf of RPs.

Since the change of target pitch angle can reduce the recognition quality, we studied its influence on the recognition quality. We found it to be very small (as illustrated in Figure 4.21) where the probability of error in the target type recognition versus target's pitch angle is shown.

A very serious factor that can distort the recognition quality is sensitivity of the ANN to the range shift of RPs within the range gate. This sensitivity

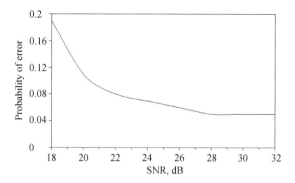

Figure 4.20 Probability of error in neural recognition of 11 target types versus SNR in dB for aspect sectors in heading of 0° to 10°, in pitch angle of −2° to 2°, and in roll angle of −3° to 3°, and for exactly known range.

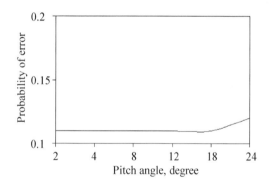

Figure 4.21 Probability of error in neural recognition of 11 target types versus target's pitch angle for SNR of 20 dB for the ANN trained as in the case of Figure 4.20.

for the ANN trained under the condition of unchanged positions of RPs in the range gate is illustrated in Figure 4.22. Here, the error probability in target type recognition versus the range shift of RPs is shown for the SNR of 20 dB. It can be seen that even a small range shift significantly increases the error probability in this case. Apparently, the influence of the range shift can be reduced by the use of the ANN with a significantly increased number of hidden nodes trained by RPs with different shifts within the range gate.

Problem of Type Recognition by Means of 2D Images. The 2D images and their shadowing effects are changed depending on orientation of each target in space. The problem consists, therefore, of too large a number of ANN hidden nodes required for recognition. The study of this subject was first

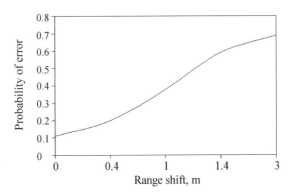

Figure 4.22 Probability of error in recognition of 11 target types versus range shift of RPs for SNR of 20 dB for the ANN trained as for the case of Figure 4.20.

carried out by Farhat in [22], who recognized scaled models of the B-52 aircraft and the Space Shuttle in an anechoic chamber. Recognition was carried out by means of an ANN under conditions of number M of RPs constituting 10% of the whole training statistics necessary to obtain the 2D image. Assuming this statistic to be equal, as usual, to $T \approx 64$ to 128 RPs, the latter can mean that the number of standard RPs for recognition was about $\Gamma \approx 6$ to 12. As one of the recognition variants, Farhat proposed to consider ANN recognition using only a single RP ($\Gamma = 1$). As we saw above, the use of a single RP ($\Gamma = 1$) allows us, in principle, to recognize at least 11 target types. We can expect, therefore, that coherent and noncoherent processing of a small number $\Gamma > 1$ of RPs will be sufficient to realize robust recognition.

4.4.5 Some Conclusions from Simulation of Neural Algorithms with Gradient Training

The simulation examples considered in Sections 4.4.3 and 4.4.4 have shown that neural recognition algorithms:

- Are robust to many distorting factors accounted for in the training. We saw above that these factors can be various aspect, range, and SNR ambiguities. The change of the aircraft structure can be another factor accounted for in the training;

- Provide recognition of target classes and types with quality indices close to those of optimal recognition algorithms;

- Do not require much time for the recognition;

- Can be burdensome in the training phase if too many factors have to be accounted for; consequently, new variants of training are of interest.

4.4.6 Perspectives of Evolutionary (Genetic) Training

Evolutionary training (ET) can be used to optimize structures of FANN NM [23, 24] and FANN M [20]. As it was for the gradient procedures, the initial set of vector $\boldsymbol{\alpha}_k = \left\| \alpha_{k1} \quad \alpha_{k2} \quad \ldots \quad \alpha_{ks} \quad \ldots \quad \beta_{ks} \quad \ldots \right\|^T$, $k = 0$ components, called "solution" or "gene," can be chosen arbitrarily, from previous experience or from other algorithms.

In a like manner, a set of such solutions $\boldsymbol{\alpha}_0^{(\xi)}$, $\xi = 1, 2, \ldots, \Xi$ large enough, known as "population," can also be introduced. The quality of each solution is then estimated using the quadratic mean cost function (4.36).

A part $\Delta\Xi_w$ of the worst solutions, providing maximum values of the mean cost function, is replaced by the new ones. For this purpose, a part $\Delta\Xi_b$ of the best solutions, providing minimum values of the mean cost function, is also arbitrarily selected to produce an "offspring," which replaces the worst solutions. After estimating the offspring quality, the described process of selection is recursively repeated. To terminate the selection process, some external criterion, such as the prescribed number of recursions or suitable error level, is applied.

The procedure of producing the offspring is called "breeding." Various kinds of breeding can be used in ET. In two-parent breeding, the two genes corresponding to "father" and "mother" can be arbitrarily chosen and combined in a certain manner. One-parent breeding can also be used where various parts of a single gene are arbitrarily combined. Three-parent breeding is recommended in [24], even though it is not a biological norm.

References

[1]   Van Trees, H. L., *Detection, Estimation and Modulation Theory*, Part 1, New York: Academic Press, 1990.

[2]   Duda, R. O., and P. E. Hart, *Pattern Classification and Scene Analysis*, New York: Wiley, 1973.

[3]   Fukunaga, K., *Introduction in Statistical Pattern Recognition*, New York: Wiley, 1990.

[4]   Patric, E. A., *Fundamentals of Pattern Recognition*, Englewood Cliffs, NJ: Prentice-Hall, 1972.

[5]   Repin, V. G., and G. P. Tartakovsky, *Statistical Syntheses by A Priori Uncertainty and Adaptation of the Information Systems*, Moscow: Sovetskoe Radio Publishing House, 1977 (in Russian).

[6]   Shirman, Y. D., "About some Algorithms of Object Classification by a Set of Features," *Radiotekhnika i Electronika*, Vol. 40, July 1995, pp. 1095–1102 (in Russian).

[7]   Shirman, Y. D. et al., "Methods of Radar Recognition and Their Simulation," *Zarubeghnaya Radioelectronika—Uspehi Sovremennoi Radioelectroniki*, No. 11, November 1996, Moscow, pp. 3–63 (in Russian).

[8]   Shirman, Y. D., S. P. Leshenko, and V. M. Orlenko, "Aerial Target Backscattering Simulations and Their Use in Technique of Radar Recognition," Vestnik Moskovskogo Gosudarstvennogo Tehnicheskogo Universiteta imeni N.E. Baumana, *Radioelectronika*, 1998, No. 4, pp. 14–25 (in Russian).

[9]   Shirman, Y. D., and S. A. Gorshkov, "Classification in Active Radar with Passive Response and in Passive Radar." In *Handbook: Electronic Systems: Construction Foundations and Theory*, Shirman, Y. D. (ed.), Moscow: Makvis, 1998, pp. 668–688, Section 24.9 (in Russian).

[10]   Gorelik, A. L. (ed.), *Selection and Recognition on the Radar Information Base*, Moscow: Radio i Svyaz Publishing House, 1990 (in Russian).

[11]   Zhuravlev, Yu. I. (ed.), *Recognition, Classification, Prediction*, Moscow: Nauka Publishing House, 1989 (in Russian).

[12]   Jouny, I., F. D. Garber, and S. Ahalt, "Classification of Radar Targets Using Synthetic Neural Networks," *IEEE Trans.*, AES-29, April 1993, pp. 336–344.

[13]   Hudson, S., and D. Psaltis, "Correlation Filters for Aircraft Identification From Radar Range Profiles," *IEEE Trans.*, AES-29, July 1993, pp. 741–748.

[14]   *Proc. IEEE*, Vol. 54, April 1996 (thematic issue about wavelets).

[15]   Rothwell, E. J. et al., "A Radar Target Discrimination Scheme Using the Discrete Wavelet Transform for Reduced Data Storage," *IEEE Trans.*, AP-42, June 1994, pp. 1033–1037.

[16]   "Model of Sampled Signal and Wavelet Signal Model." In *Handbook: Electronic Systems: Construction Foundations and Theory*, Second edition, Section 13.6, Kharkov (printing in Russian).

[17]   Zurada, J., *Introduction to Artificial Neural Systems*, West Publishing Com., 1992.

[18]   Rummelhart, D., and T. McClelland (eds.), *Parallel Distributed Processing*, Cambridge, MA: MIT Press, 1988.

[19]   Werbos, P., "Backpropagation Through Time," *Proc. IEEE*, Vol. 78, October 1990, pp. 1550–1560.

[20]   Caelly, T., L. Guan, and W. Wen, "Modularity in Neural Computing," *Proc. IEEE*, Vol. 87, September 1999, pp. 1496–1518.

[21]   Orlenko, V. M., and Y. D. Shirman, "Radar Target's Neural Recognition Accounting for Distorting Factors," *Collection of Papers*, Issue 3, 2000, Moscow: Radiotekhnika Publishing House, pp. 82–85 (in Russian).

[22]   Farhat, N. H., "Microwave Imaging and Automated Target Identification Based on Models of Neural Networks," *Proc. IEEE*, Vol. 77, May 1989.

[23]   Fogel, D. B., T. Fukuda, and L. Guan, "Technology on Computational Intelligence," *Proc. IEEE*, Vol. 87, September 1999, pp. 1415–1422.

[24]   Yao, X., "Evolving Artificial Neural Networks," *Proc. IEEE*, Vol. 87, September 1999, pp. 1423–1470.

[25]   Libby, E. W., and P. S. Maybeck, "Sequence Comparison Techniques for Multisensor Data Fusion and Target Recognition," *IEEE Trans.*, AES-32, No 1, January 1996, pp. 52–65.

[26]   Jacobs S. P., and J. A. O'Sullivan, "Automatic Target Recognition Using Sequences of High Resolution Radar Range Profiles," *IEEE Trans.*, AES-36, April 2000, pp. 364–383.

5

Peculiarity of Backscattering Simulation and Recognition for Low-Altitude Targets

Essential factors to be simulated in the case of low-altitude targets are the surface reflections (1) of the transmitted signal and (2) of the signal backscattered by the target. The first kind of reflection produces the ground clutter. The second kind of reflection, together with the first and distortions in a clutter rejection device, changes the amplitude of the received narrowband signal and distorts the structure of the received wideband signal. Simulation of ground clutter is discussed in Section 5.1. Simulation of signal distortions caused by the ground reflections from a ground or sea surface is considered in Section 5.2. The problem of the target wideband recognition under conditions of signal distortions and clutter is stated and discussed in Section 5.3.

5.1 Ground Clutter Simulation

Creation of electrodynamic models of backscattering from the Earth's surface at centimeter and decimeter wavelengths is encumbered by its complexity and the diversity of its types [1, 2]. Empirical models are therefore preferred. We consider basic parameters of empirical simulation in Section 5.1.1. Calculation of the complex amplitude of clutter necessary for simulation is carried out in Section 5.1.2. In Sections 5.1.3 and 5.1.4 we consider clutter simulation in more detail, using the digital terrain maps (DTM).

5.1.1 Basic Parameters of Empirical Simulation

Basic parameters of the empirical simulation are the specific RCS (m^2/m^2) and specific PSM (m^2/m^2) for the backscattering from the ground and from objects on the ground, their statistical distributions, and the power spectra of their fluctuations.

Empirical Description of Specific RCS. An example of such a description is given in [3, 4] as a function of frequency in the range f = 3 to 100 GHz, of grazing angle $\psi \le 30°$, and of the surface parameters A_1, A_2, A_3 (Table 5.1):

$$\sigma_0(f, \psi),\ dB = A_1 + A_2 \lg\frac{\psi}{20} + A_3 \lg\frac{f}{10} \qquad (5.1)$$

where ψ is given in degrees, and f in GHz.

Complementary to Table 5.1 it should be noted that the specific RCS of a forest is increased by 5 dB after the summer rain. The specific RCS of most surfaces is greater by 2 to 3 dB for vertical than for horizontal polarization. The ratio of specific RCS values observed at matched (intended) and cross-polarization is about 10 to 15 dB.

The data in Table 5.1 will be used in simulation for vertical polarization directly and for horizontal and cross-polarizations with the noted correction.

Table 5.1
Parameters of Surfaces of Various Types

Surface Type/Surface Parameters	A_1	A_2	A_3
	(m^2/m^2)	(m^2/m^2)	(m^2/m^2)
Concrete	−49	32	20
Plough-land	−37	18	15
Snow	−34	25	15
Leafy forest, summer	−20	10	6
Leafy forest, winter	−40	10	6
Pine forest, summer and winter	−20	10	6
Meadow with the grass of more than 0.5m height	−21	10	6
Meadow with the grass of less than 0.5m height	−28	10	6
Town and country buildings	−8.5	5	3

Statistical Distribution of Specific RCS.   The specific RCS of a ground surface is a random function of three parameters: of two coordinates and of time. The resultant statistical distribution is frequently described by the log-normal, Weibull, K, and other non-Gaussian models. Such an approach to the coordinate distributions does not allow accounting for the variety of Earth surface relief and cover. Digital terrain maps will therefore be used allowing us to determine the distinctive surface elements (see Section 5.1.2). As for the time dependencies of the quadrature components of reflections from local surface elements, they can be described by Gaussian stationary random functions with a definite power spectrum of fluctuations.

Power Spectrum of Fluctuations and Its Use in Simulation.   The power spectrum of specific RCS fluctuations is formed by two terms corresponding to the part of RCS that is stable in time (building, rock, surface without vegetation, etc.) and to the part of RCS varying in time (surface with vegetation). This can be approximated in the following form:

$$G(F) \approx G_0 \left\{ a^2 \delta(F) + k\left[1 + \left(\frac{F}{\Delta F}\right)^n \right]^{-1} \right\}, \; k = \left\{ \int_0^\infty \left[1 + \left(\frac{F}{\Delta F}\right)^n \right]^{-1} df \right\}^{-1}$$

$$(5.2)$$

where G_0 is the spectral density at zero frequency and $a^2 = C(\overline{V})^{-m}$ is the ratio of the stable part of specific RCS to the mean value of the varying one depending on the mean wind velocity $\overline{V}$ near the surface and the characteristics of the surface. Here, $C \approx 10^4$ to 10^5, $m = 0$ for town and country buildings, $C \approx 300$, $m \approx 2.9$ for a forest, $C \approx 60\lambda^2$, $m \approx 3$ for scrubby vegetation, $C \approx 100\lambda$, $m \approx 3$ for a surface free of woods in winter; $\Delta F \cong 0.04 \cdot \overline{V}^{1.3}/\lambda$ is the spectrum width at the 3-dB level; $n = 2(\overline{V} + 2)/(\overline{V} + 1)$ [3, 4].

The approximate expression (5.2) can be used in simulation. Let us form a complex realization $\dot{Y}(t)$ of the random stationary process with power spectrum $[1 + (F/\Delta F)^n]^{-1}$ with zero mean and unity variance from white noise by analogy with Figure 1.10. Let's add to it the stable part with mean a and random initial phase φ distributed uniformly over the surface resolution cells. The simple approximation of the normalized process with the spectrum (5.2) will be

$$\dot{Y}_a(t) \approx [\dot{Y}(t) + a \cdot e^{j\varphi}]/\sqrt{1 + a^2} \qquad (5.3)$$

Specific Polarization Backscattering Matrix of Ground Surface. The specific PSM $\mathbf{B}_{l,n}(f, \psi)$ will be used below for calculating backscattering from various ground surface elements l, n ($l = 1, 2, \ldots$; $n = 1, 2, \ldots$), at various frequencies f, and grazing angles $\psi = \psi_{l,n}$ observed with vertical (V) and horizontal (H) polarizations:

$$\mathbf{B}_{l,n}(f, \psi) = \left\| \begin{array}{cc} \sqrt{\sigma_{l,n}^{VV}(f, \psi)} & \sqrt{\sigma_{l,n}^{VH}(f, \psi)} \\ \sqrt{\sigma_{l,n}^{HV}(f, \psi)} & \sqrt{\sigma_{l,n}^{HH}(f, \psi)} \end{array} \right\| \qquad (5.4)$$

The data on specific PSM elements necessary for simulation can be obtained from (5.1), Table 5.1, and notes related to them. Decomposition of the surface into the l, n elements is clarified in Figure 5.1. The number l corresponds to the sample of the horizon range $r_{hl} = l\delta r$, where δr is the interval of the range sampling that is equal usually to the interval of range resolution. The number n corresponds to the sample of azimuth $\beta_n = n\delta\beta$, where $\delta\beta$ is the interval of azimuth sampling. The value $\delta\beta$ in radians is frequently chosen in the form $\delta\beta = \Omega_a T$, where Ω_a is the angular rate of antenna beam rotation in radians/s, and T is the pulse repetition interval in seconds. Let us agree to consider the horizon range r_{hl} as a distance along the horizon plane taking into account the Earth's curvature and refraction in the reference atmosphere (see also Section 5.1.2).

5.1.2 Calculation of the Clutter Complex Amplitude

The integrated polarization scattering matrix $\mathbf{D}_l(k)$ is calculated for the lth integrated range resolution element of the surface within the antenna pattern $F(\beta, \epsilon)$. This pattern is oriented in direction $\beta_k = k\delta\beta$ and receives clutter from the azimuthal coverage sector $(k - M)\delta\beta \le \beta \le (k + M)\delta\beta$ (Figure 5.1). Then,

$$\mathbf{D}_l(k) = \sqrt{\delta S_l} \sum_{n=k-M}^{k+M} F^2[(n-k)\delta\beta, \epsilon_{l,n}] \mathbf{B}_{l,n}(f, \psi_{l,n}) e^{-j2\pi f\delta t_{l,n}}$$

$$(5.5)$$

Here,

$\delta S_l = \delta r_l \cdot r_{hl}\delta\beta$ is the area of the l, nth element of the surface at the horizon range r_{hl} [the square root of $\delta S_{l,n}$ appears through properties of (5.4)];

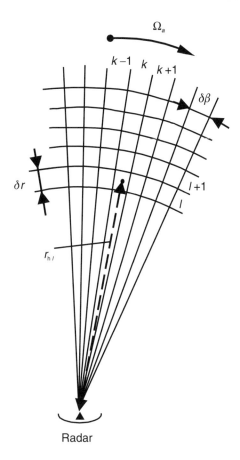

Figure 5.1 Decomposition of the surface within the coverage sector and range resolution intervals into elements.

$F(\beta, \epsilon) = \left| \dot{F}(\beta, \epsilon) \right|$ is the transmission-reception pattern of the radar antenna;

$\epsilon_{l,n}$ is the elevation angle of the l, nth surface element

$$\epsilon_{l,n} = \operatorname{atan}[(y_a - y_{l,n})/r_{hl}] \tag{5.6}$$

where $y_a - y_{l,n}$ is the difference of heights between the radar antenna and the l, nth element of surface; $\psi_{l,n}$ is the grazing angle of the l, nth surface element

$$\psi_{l,n} = \epsilon_{l,n} + \operatorname{atan}[\partial y_{l,n}/\partial r_h] - r_{hl}/R_{ef} \tag{5.7}$$

R_{ef} is the effective Earth radius accounting for the atmosphere refraction of radio waves; and

$\delta t_{l,n} = [2r_l(1/\cos(\epsilon_{l,n}) - 1)]/c$ is the increment of delay time from the l, nth surface element due to nonzero elevation angle $\epsilon_{l,n}$ and slant range r_l.

The sum (5.5) includes only the terms corresponding to illuminated areas of the Earth's surface (i.e., the areas for which the inequality $\epsilon_{l,n} > \epsilon_{q,n}$, $q = 1, \ldots, l - 1$ is valid). Here, l and q are the numbers of surface elements increasing with their horizon range r_{hq}.

The final expression of the clutter complex amplitude is found similarly to expression (1.20), given in Section 1.3.4. Its value in units of noise standard deviation is

$$\dot{E}(t, \beta_k) = (\mathbf{p}_{rec}^0)^{*T}\left\{\sum_{l=1}^{N} \dot{Y}_{al}(t)\mathbf{D}_l(k)e^{-j2\pi f \Delta t_l}\dot{U}(t - \Delta t_l)\, 10^{-Q_l/20}\sqrt{W}/r_l^2\right\}\mathbf{p}_{tr}^0$$

(5.8)

Here,

$\mathbf{p}_{rec}^0$ and $\mathbf{p}_{tr}^0$ are the polarization vectors of receiving and transmitting antennas;

$\dot{Y}_{al}(t) = (\dot{Y}_l(t) + a \cdot e^{j\varphi_l})/\sqrt{1 + a^2}$ is a random complex multiplier of the type (5.3) corresponding to the lth integrated surface element (the possibility of the reflection depolarization here is not accounted for);

$\Delta t_l = 2r_{hl}/c$ is the delay of the echo from the lth integrated surface element;

W is the radar potential (3.2) in m^2 ensuring normalization of the $E(t, \beta_k)$ value in relation to the noise standard deviation; and

Q_l is the factor for additional losses in dB due to propagation.

5.1.3   Use of Digital Terrain Maps in Simulation

In this section we shall consider (1) general knowledge of DTM, (2) DTM information used in ground clutter simulation, (3) stages of ground clutter simulation, (4) microrelief simulation, (5) the influence of the Earth's surface curvature and of atmosphere refraction, and (6) examples of ground clutter simulation.

5.1.3.1   General Knowledge of DTM

DTMs are widely used [5, 6]. These contain information about the terrain space distribution and replace the non-Gaussian statistical models (lognormal, Weibull, and K types) widely used earlier. Up-to-date DTMs correspond to databases that include information on terrain relief, objects giving an increase of relative height, roads, hydrographic objects, and types of surface covering. The accuracy of cartographic information increases systematically by means of geo-informational space systems, for instance. But as yet the accuracy of commercial DTMs does not exceed 5% to 10% of the map scale unit, being limited by coverage of large regions and restricted computational capabilities.

5.1.3.2   DTM Information Used in Ground Clutter Simulation

Such information consists of:

- Absolute heights of the terrain macrorelief given in matrix form with steps of 50 to 100m and with indication of a surface type;
- Relative heights of forests, buildings, etc.;
- Additional data about various manmade and natural point objects (masts, towers, summits of mountains and hills), hydrographic objects (rivers, lakes, marshes, seas), and features of transport systems (including bridges).

Seasonal information (summer, winter, presence of snow) and current weather conditions are provided as well.

Unfortunately the variety of DTM formats does not allow for providing a general formula for extracting the necessary information. The authors' proposition is to specify the format of the data necessary for simulation. This will allow the users to order the needed format converters to the special-purpose geo-informational firms.

5.1.3.3   Stages of Ground Clutter Simulation

This simulation can be divided into three stages. After inputting the data on the radar type, season, and weather conditions (first preliminary stage), auxiliary calculations are provided (second stage). The third (executive) stage is the simulation of ground clutter. The subjects of the first, second, and third stages are clarified in the block diagrams (Figure 5.2).

5.1.3.4   Microrelief Simulation

This type of simulation considers the surface roughness with characteristic dimensions that are essential for high-resolution radar but are not considered

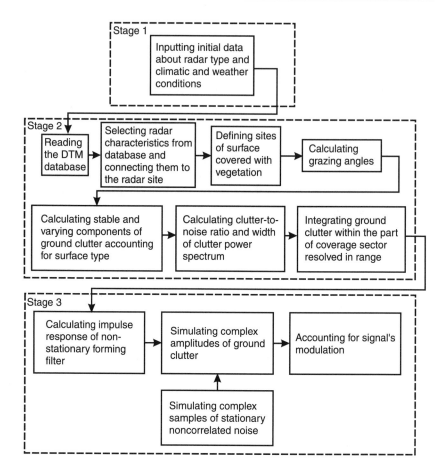

Figure 5.2 Subjects of the first, second, and third stages of ground clutter simulation.

by DTM. Relative heights of the microrelief elements observed in a particular direction correspond to realization of a stationary random process $f(l)$ with zero mean and variance $\sigma_y^2 (\text{m}^2)$. Realizations of this random process can be defined by an empirical "two-scale" correlation function [7]:

$$r_y(l) = \sigma_y^2 \cdot [A_1 \exp(-\alpha_1|l|) \cdot \cos(\beta_1|l|) + A_2 \exp(-\alpha_2|l|) \cdot \cos(\beta_2|l|)]$$
$$(5.9)$$

where values α_i and β_i are the correlation parameters, and values A_1 $(0 \leq A_1 \leq 1)$ and $A_2 = 1 - A_1$ are the corresponding dimensionless coefficients for the two scales. All these values are presented in Table 5.2 for various underlying surfaces [7, 8].

Table 5.2
Microrelief Parameters for Various Underlying Surfaces

No.	Type of Underlying Surfaces	Variant of the Surface	σ_y (cm)	A_1	α_1 (cm^{-1})	β_1 (cm^{-1})	A_2	α_2 (cm^{-1})	β_2 (cm^{-1})
1	Artificial asphalt carpet	I	1.5	1.0	0.15	0.0	0.0	0.0	0.0
		II	1.0	1.0	0.15	0.0	0.0	0.0	0.0
		III	0.1	1.0	0.2	0.0	0.0	0.0	0.0
2	Stony terrain	I	2.5	1.0	0.45	0.0	0.0	0.0	0.0
		II	5	0.9	0.2	0.0	0.1	0.05	1.4
		III	10	0.8	0.3	0.0	0.2	0.2	1.7
3	Rugged terrain (slightly to intensely rugged)	I	10.0	1.0	0.2	0.6	0.0	0.0	0.0
		II	15.0	1.0	0.12	0.3	0.0	0.0	0.0
		III	20.0	1.0	0.1	0.2	0.0	0.0	0.0
4	Plough-land across the furrows	I	4	0.9	0.5	0.0	0.1	0.4	6.5
		II	15	1.0	0.06	0.08	0.0	0.0	0.0

5.1.3.5 Influence of the Earth's Surface Curvature and Atmosphere Refraction

For each lth ($l = 1, 2, \ldots, N$) surface backscattering element with coordinates x_{Dl}, y_{Dl}, z_{Dl} relative to the point 0, y_a, 0 of the radar antenna location, one can find its horizon range to the radar without accounting for the Earth's curvature and refraction as $r_{Dl} = \sqrt{x_{Dl}^2 + z_{Dl}^2}$ and its height relative to radar antenna as $y_{Dl} - y_a$ (Figure 5.3). Here, y_{Dl} and y_a are the DTM heights of the lth surface element and of the radar antenna.

The height y_l of the lth integrated surface element beyond the radio horizon accounting for atmosphere refraction can be found using the approximate expression $\sqrt{1 + \alpha} \approx 1 + \alpha/2$ for $|\alpha| = (r_{hl}/R_{ef})^2 \ll 1$. Then we have

$$y_l = y_{Dl} + R_{ef} - \sqrt{R_{ef}^2 + r_{hl}^2} \approx y_{Dl} - r_{hl}^2 / 2R_{ef} \qquad (5.10)$$

Here, R_{ef} is the effective Earth radius accounting for refraction in the reference atmosphere (8.5×10^6 m), and r_{hl} is the corresponding horizon range (distance along the horizon plane). Solving the right triangle, using

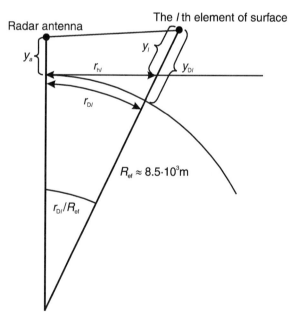

Figure 5.3 Clarification of the clutter simulation geometry accounting for the Earth's curvature and refraction in the reference atmosphere.

the approximate expression $\sin\varphi \approx \varphi$ for $|\varphi| = r_{Dl}/R_{ef} \ll 1$ and taking into account (5.10), we obtain with accuracy up to inverse second power of the effective Earth radius (R_{ef}^{-2})

$$r_{hl} = (R_{ef} + y_{Dl} - y_l)\sin(r_{Dl}/R_{ef}) \approx r_{Dl}[1 + r_{hl}^2/2R_{ef}^2] \approx r_{Dl} \tag{5.11}$$

5.1.3.6 Example of Ground Clutter Simulation

An example of the ground clutter simulation is presented in Figure 5.4(a) and (b). Here, the matrix of absolute heights, obtained from DTM data, is shown [Figure 5.4(a)] for the Eastern European subdistrict with dimension 50×50 km; the brighter parts of illustration correspond to the greater heights, and the darkest parts correspond to lowlands. The result of integration of ground clutter in the angle coverage sector is shown in Figure 5.4(b). The simulated view of the plan-position indicator is shown in Figure 5.4(c) as a result of modeling the target and the ground clutter from DTM data.

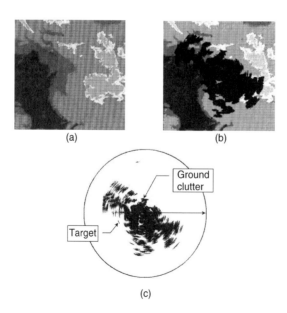

(a)   (b)

Ground clutter

Target

(c)

Figure 5.4 Simulation of ground clutter: (a) DTM matrix of heights in the brightness description for Eastern European, 50×50 km subdistrict (greater heights are brighter); (b) result of clutter integration in a coverage sector (clutter is blackened); and (c) simulated PPI view (clutter is blackened).

5.2 Simulation of Distortions of Signal Amplitude and Structure

In this section we consider the principles of simulation (Section 5.2.1) and approximate solutions of the scattering problem at the Earth-atmosphere interface (Sections 5.2.2 and 5.2.3) using initial data obtained from the DTM. The main factors contributing to wave propagation above an underlying surface are analyzed in Section 5.2.4. In Section 5.2.5, on the basis of simulation, we investigate the influence of surface reflections on amplitude and structure of wideband and narrowband signals.

5.2.1 Principles of Simulation of Wave Propagation Above Underlying Surface

The DTM, containing the necessary a priori information about the macrorelief and the surface type in the target direction, is most convenient for simulation of surface reflections. We also need additional information about the microrelief, electrical parameters of the underlying surface (complex dielectric constant, conductivity), and current weather and atmosphere conditions (a presence of atmospheric precipitates, their type and intensity, temperature, atmospheric pressure, humidity). Discrete description of the underlying surface can be supplemented by the preliminary continuous one, using, for instance, a spline approximation. Smoothed descriptions are used to define the first and second derivatives in the surrounding of points of ray reflection.

Principles of Approximate Solution of the Scattering Problem. The approximate solution is carried out on the basis of geometrical optics [9–11] and the theory of multipath propagation [1–3, 10, 11]. It is assumed that the uneven but smooth surface (Figure 5.5) contains a great number M of reflecting elements. Each of them is a source of a spherical wave and contains several Fresnel zones and points of stationary phase.

The resultant field $\dot{E}_\Sigma$ near the receiving antenna is defined as the sum of the $1 + 2M + M^2$ backscattered ray groups

$$\dot{E}_\Sigma = \dot{E}_{00} + \sum_{m=1}^{M} \dot{E}_{m0} + \sum_{l=1}^{M} \dot{E}_{0l} + \sum_{m=1}^{M}\sum_{l=1}^{M} \dot{E}_{ml} \qquad (5.12)$$

of four kinds:

1. $\dot{E}_{00}$ corresponding to direct target illumination and direct reception without reflections from the surface elements;

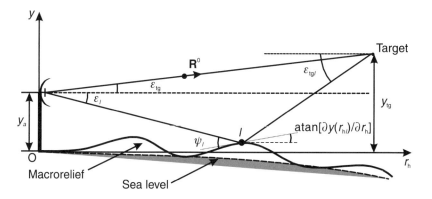

Figure 5.5 Clarification of radar signal formation near an uneven smooth surface.

2. $\dot{E}_{m0}$, k = 1, 2, ..., M corresponding to illumination of the target through the mth surface element (Section 5.2.1) and direct reception;

3. $\dot{E}_{0l}$, l = 1, 2, ..., M corresponding to direct target illumination and reception through the lth surface element;

4. $\dot{E}_{ml}$, k, l = 1, 2, ..., M corresponding to target illumination through the mth surface element and reception through the lth surface element.

Conditions of Applicability. The four following assumptions are made:

- The target range exceeds the range of masking (shadowing) obstacles restricting direct visibility of the target;

- Grazing angles ψ_l (Figure 5.5) are less than 30° and more than 0.5° to 1°, allowing use of the results of [1–3, 9–12]. For the angles less than 0.5° to 1° it is possible also to use the results of [9] obtained for the shadow and half-shadow zones;

- For small grazing angles the monostatic PSM $\mathbf{A}_i(\mathbf{R}°)$ replaces approximately the bistatic PSM $\mathbf{A}_i(\mathbf{R}°, \mathbf{L}_{l,i})$ for each lth bright element of the surface and ith bright element of the target. Although the values $|\mathbf{L}|$ are not small, moduli of the dot products $|\mathbf{L}^T dE/d\mathbf{L}|$ entered in (1.21) are not usually very large;

- Path-length differences between the direct rays and those propagated through arbitrary lth or mth (l, m = 1, ..., M) element of surface (Figure 5.5) are approximately identical for various bright elements of target i (i = 1, ..., N).

5.2.2　Approximate Solution of the Scattering Problem at the Earth-Atmosphere Interface

Components of (5.12), after accounting for the conditions discussed above, and angular resolution of surface elements, will take the following form:

$$
\dot{E}_{00} = (\mathbf{p}_{\mathrm{rec}}^0)^{*\mathrm{T}} \left[\sum_{i=1}^{N} \dot{F}_{\mathrm{rec}}(\beta_i, \epsilon_i) \mathbf{A}_i(\mathbf{R}^\circ) \dot{F}_{\mathrm{tr}}(\beta_i, \epsilon_i) e^{-j2\pi f \Delta t_i} \dot{U}(t - \Delta t_i) \right] \mathbf{p}_{\mathrm{tr}}^0
$$

(5.13)

$$
\dot{E}_{m0} = (\mathbf{p}_{\mathrm{rec}}^0)^{*\mathrm{T}}
$$
$$
\cdot \left[\sum_{i=1}^{N} \dot{F}_{\mathrm{rec}}(\beta_i, \epsilon_i) \mathbf{A}_i(\mathbf{R}^\circ) e^{-j2\pi f(\Delta t_i + \delta t_m)} \dot{U}(t - \Delta t_i - \delta t_m) \right]
$$
$$
\cdot \mathbf{B}_m \mathbf{p}_{\mathrm{tr}}^0 \dot{F}_{\mathrm{tr}}(\beta_m, \epsilon_m)
$$

(5.14)

$$
\dot{E}_{0l} = \dot{F}_{\mathrm{rec}}(\beta_l, \epsilon_l)(\mathbf{p}_{\mathrm{rec}}^0)^{*\mathrm{T}} \mathbf{B}_l^{*\mathrm{T}}
$$

(5.15)

$$
\cdot \left[\sum_{i=1}^{N} \mathbf{A}_i(\mathbf{R}^\circ) \dot{F}_{\mathrm{tr}}(\beta_i, \epsilon_i) e^{-j2\pi f(\Delta t_i + \delta t_l)} \dot{U}(t - \Delta t_i - \delta t_l) \right] \mathbf{p}_{\mathrm{tr}}^0
$$

$$
\dot{E}_{kl} = \dot{F}_{\mathrm{rec}}(\beta_l, \epsilon_l)(\mathbf{p}_{\mathrm{rec}}^0)^{*\mathrm{T}} \mathbf{B}_l^{*\mathrm{T}}
$$

(5.16)

$$
\cdot \left[\sum_{i=1}^{N} \mathbf{A}_i(\mathbf{R}^\circ) e^{-j2\pi f(\Delta t_i + \delta t_{kl})} \dot{U}(t - \Delta t_i - \delta t_{ml}) \right] \mathbf{B}_m \mathbf{p}_{\mathrm{tr}}^0 \dot{F}_{\mathrm{tr}}(\beta_m, \epsilon_m)
$$

$$
\delta t_{ml} = \delta t_m + \delta t_l
$$

(5.17)

Here,

$\dot{F}_{\mathrm{tr}}(\beta, \epsilon)$ and $\dot{F}_{\mathrm{rec}}(\beta, \epsilon)$ are the complex patterns of the transmitting and receiving antennas;

$\Delta t_i = 2r_i/c$ is the direct echo delay from the ith target element;

δt_m and δt_l are the additional delays due to the ray propagation through the mth surface element on the "radar-target" path and through the lth surface element on the "target-radar" path; and

$\mathbf{B}_m$ and $\mathbf{B}_l$ are the specific PSMs for the mth and lth surface elements.

5.2.3 Variants of Approximate Solutions of the Scattering Problem

Use of Operational Procedure for Directly Obtaining the Result of (5.13) to (5.16). The operational form of the described solution (5.13) to (5.16) is a development of (1.22) through (1.23):

$$A(p) = \mathbf{P}_{rec}^{*T}(p) \left[\sum_{i=1}^{N} \Lambda_i (\mathbf{R}^\circ, \mathbf{L}) e^{-pt_i} 10^{-Q_{Abi}/20} \right] \mathbf{P}_{tr}(p) \qquad (5.18)$$

where $\mathbf{P}_{tr}(p)$ and $\mathbf{P}_{rec}(p)$ are the operational forms of the propagation-polarization vectors for the paths "transmitter-target" and "target-receiver":

$$\mathbf{P}_{tr}(p) = \left(\mathbf{I} + \sum_{l=1}^{M} \dot{F}_{tr}(\beta_l, \epsilon_l) \mathbf{B}_l e^{-p\delta t_l} \right) \cdot \mathbf{p}_{tr}^0 \qquad (5.19)$$

$$\mathbf{P}_{rec}(p) = \left(\mathbf{I} + \sum_{l=1}^{M} \dot{F}_{rec}(\beta_l, \epsilon_l) \mathbf{B}_l e^{-p\delta t_l} \right) \cdot \mathbf{p}_{rec}^0$$

Performing the inverse transforms (1.22), we can obtain the result of substitution of equations (5.13) through (5.17) into (5.12).

Use of Facet Procedure. The facet procedure is realized by means of replacing a smooth surface by a surface consisting of contiguous triangles (see Section 7.1.5). As with the previous procedures, this one can be based on the use of DTM information.

5.2.4 Main Factors Contributing to the Wave Propagation Above Underlying Surface

Description of relief, geometrical parameters of reflections, and electrical parameters of underlying surface contribute to the wave propagation above this surface.

Description of relief is provided by:

1. Supplementing the discrete description of the underlying surface with the smoothed one $y_l = y(r_{hl})$, where r_{hl} is the horizon range (5.11);
2. Using also the microrelief correlation function (5.9).

Geometrical parameters of reflections include the reflection-effective domain, coordinates of reflecting elements, and path-length differences.

The reflection-effective domain is the region significant for reflection. It is determined by the total dimensions of several main Fresnel zones. All these Fresnel zones are disposed in the vicinities of various kth bright points ($k = 1, \ldots, M$) and have elliptical shapes [9–11]. The length (radial range extension) and width (cross-range extension) of each zone is limited by the path-length difference $n\lambda/4$ of the beams reflected from its edge points. The length of the total ellipse is hundreds of meters or kilometers. It significantly exceeds the width, which does not, as a rule, exceed meters or dozens of meters [3, 9–11]. The roughness of relief in the cross-range direction can therefore be neglected if the country is not very hilly.

Coordinates of Reflecting Points and Checking for Their Visibility. The coordinates of reflecting elements $r_{\mathrm{h}l}$ and $y(r_{\mathrm{h}l})$ for the rays that reach the target (Figure 5.5) can be found from the transcendental equation

$$\epsilon_l + \mathrm{atan}[\partial y(r_{\mathrm{h}l})/\partial r_{\mathrm{h}}] = \epsilon_{\mathrm{tg}l} - \mathrm{atan}[\partial y(r_{\mathrm{h}l})/\partial r_{\mathrm{h}}] \qquad (5.20)$$

where ϵ_l is the elevation angle of the lth surface element from the radar side, and ϵ_{tg} is the elevation angle of this element from the target side. Both angles are calculated using additional equations analogous to (5.6).

The conditions of visibility of lth reflecting element from both the radar antenna's phase center and target position must be checked against

$$\begin{cases} \tan\epsilon_l > \tan\epsilon_q, \text{ for } q = 1, \ldots, l-1 \\ \tan\epsilon_{\mathrm{tg}l} > \tan\epsilon_{\mathrm{tg}q}, \text{ for } q = l+1, \ldots, Q \end{cases} \qquad (5.21)$$

Here, Q is the total number of the surface reflecting elements in the reflection-effective domain, q is the number of the element, increasing together with its horizontal range $r_{\mathrm{h}q}$. If inequalities (5.21) are not satisfied for element number l, then the corresponding lth reflecting element is considered to be shadowed and is excluded from further calculations. After all the points were checked, the number M is found for the points that are the real solutions of (5.20) for the unshadowed surface bright points, existing at the ranges from the minimum range to the range of the target or the range of the direct visibility.

Time delays δt_l between the direct ray and the corresponding rays propagated through lth ($l = 1, \ldots, M$) surface elements (Figure 5.5) are determined by the following equation:

$$\delta t_l = R(\cos\epsilon_{\mathrm{tg}} - \cos\epsilon_l)/c\cos\epsilon_l \qquad (5.22)$$

Electrical parameters of reflections include the complex permittivity and specific PSM of surface elements.

The complex permittivity for the lth reflecting point is found from

$$\dot{\epsilon}_{cl} = \epsilon_{rl} - j60\lambda\sigma_{cond\,l} \qquad (5.23)$$

where ϵ_{rl} is its relative permittivity and $\sigma_{cond\,l}$ is its conductivity (see Table 5.3).

Specific PSM of Underlying Surface. The specific PSM for lth reflecting surface element has a structure similar to that of the specific PSM (5.4) for ground surface scattering. Accounting for the phase, we obtain the specific PSM $\mathbf{B}_l$ in the form

$$\mathbf{B}_l = \left\|\begin{array}{cc} \sqrt{\sigma_l^{VV}}\,e^{j\varphi_l^{VV}} & \sqrt{\sigma_l^{VH}}\,e^{j\varphi_l^{VH}} \\ \sqrt{\sigma_l^{HV}}\,e^{j\varphi_l^{HV}} & \sqrt{\sigma_l^{HH}}\,e^{j\varphi_l^{HH}} \end{array}\right\| = \left\|\begin{array}{cc} \dot{R}_l^{VV} & \dot{R}_l^{VH} \\ \dot{R}_l^{HV} & \dot{R}_l^{HH} \end{array}\right\|$$
$$\cdot (1 - K_{SRl})(1 - K_{Abl})K_{RDl}K_{Shl} \qquad (5.24)$$

where $\dot{R}_l^{VV}, \dot{R}_l^{VH}, \dot{R}_l^{HV}, \dot{R}_l^{HH}$, are the complex coefficients of reflection for various polarizations, K_{SRl} is the coefficient of surface roughness, K_{Abl} is the coefficient of absorption ($1 - K_{Abl} = 0.3$–0.03 [1, p. 295]), K_{RDl} is the coefficient of the ray divergence, and K_{Shl} is the coefficient of shadowing (masking). All the coefficients are given for the lth reflecting point. The

Table 5.3
Electrical Properties of Typical Surfaces

No.	Material	ϵ_r	$\sigma_{cond},$ $(ohm*m)^{-1}$
1.	Good soil (wet)	25	0.02
2.	Average soil	15	0.005
3.	Poor soil (dry)	3	0.001
4.	Snow, ice	3	0.001
5.	Fresh water λ = 1m	81	0.7
	λ = 0.03m	65	15
6.	Salt water λ = 1m	75	5
	λ = 0.03m	60	15

Source: [1].

values $\dot{R}_l^{VV}$ and $\dot{R}_l^{HH}$ are found using the Fresnel equations for the unit relative permeability $\mu_r = 1$:

$$\dot{R}_l^{VV} = \frac{\dot{\epsilon}_{cl} \sin\psi_l - \sqrt{\dot{\epsilon}_{cl} - \cos^2\psi_l}}{\dot{\epsilon}_{cl} \sin\psi_l + \sqrt{\dot{\epsilon}_{cl} - \cos^2\psi_l}}, \quad \dot{R}_l^{HH} = \frac{\sin\psi_l - \sqrt{\dot{\epsilon}_{cl} - \cos^2\psi_l}}{\sin\psi_l + \sqrt{\dot{\epsilon}_{cl} - \cos^2\psi_l}}^*$$

$$(5.25)$$

As in [3, p. 18] and [13, p. 20], we assume in simulation that

$$\dot{R}_l^{VH} \approx 0, \quad \dot{R}_l^{HV} \approx 0 \qquad (5.26)$$

The rest of the values entered into (5.14) are defined as follows:

1. The coefficient of surface roughness K_{SRl} is defined from the equation

$$(1 - K_{SRl})^2 = \exp(-8\pi^2\sigma_{hl}^2 \sin^2\psi_l/\lambda^2) \qquad (5.27)$$

where σ_{hl} is the standard deviation of the heights of microrelief roughness in the neighborhood of the lth reflecting element [1, p. 293].

2. The divergence factor K_{RDl} is defined from the equation [9, 10, 12]

$$K_{RDl} \cong [1 + 4\mu_l^2(1 - \mu_l)^2 R_{tgh}^2(t)/R_l H_l]^{-1/2} \qquad (5.28)$$

where $R_{tgh}(t)$ is the horizon range of the target, $\mu_l = r_{hl}/R_{tgh}(t)$ is the relative range of the lth reflecting element, $H_l = r_{hl}\tan\epsilon_{tg} + y_a - y(r_{hl})$ is the altitude of the target line-of-sight above the lth reflecting element, and R_l is the average radius of the surface curvature in the neighborhood of lth reflecting element.

3. Coefficient of shadowing (masking) K_{Shl} is defined from the equation

$$K_{Shl} = N_{illl}/N_{ovl} \qquad (5.29)$$

---

* See also (7.9) and (7.10).

where N_{ill} is the number of the elements of microrelief illuminated by the radar per the unit area (line) of macrorelief in the neighborhood of lth reflecting element of surface, N_{ovl} is the overall number of the microrelief elements per the unit area (line) of macrorelief also in the neighborhood of this lth element.

5.2.5   The Influence of Surface Reflections on the Amplitude and Structure of Radar Signals

Such influence is caused by the interference of direct signals with those propagated via the Earth's surface. The signals reflected from the surface have additional delays δt_m, δt_l, or $\delta t_m + \delta t_l$ due to rays' propagation through the paths radar-surface-target, target-surface-radar, or radar-surface-target-surface-radar, respectively. The result of interference of the signals depends on the products of their bandwidth and the time delay, on their phase differences, and on the ratios of their amplitudes. The interference of narrowband signals causes practically only amplitude fluctuations; while the interference of wideband signals (target RPs for instance), together with amplitude fluctuations, causes distortions of their structures, especially in the case where direct and reflected signals are out of phase and the Fresnel reflection coefficients are near unity.

Quantitatively the influence of interference on the received narrowband signal is expressed through the pattern propagation factor, which is equal to the ratio of the amplitude of the total received signal to the amplitude of the received signal in free space. For a wideband signal this factor is defined as the ratio of the correlation processing result for the total received signal to that for the received signal in free space. In this section correlation processing is carried out only in regard to the expected RP.

Figure 5.6 shows the pattern propagation factors for a point target:

- For a narrowband signal with horizontal polarization, obtained by calculation using known formulae [1, p. 291, (6.2.5)] $F(\epsilon_{tg}) = \left| f(\epsilon_{tg}) + f(-\psi)\rho D \exp(-j\alpha) \right|$ (solid lines in Figure 5.6);
- For a wideband chirp signal of 300-MHz deviation with horizontal polarization, obtained by simulation using standard RPs for free space (dotted and dashed lines in Figure 5.6).

The calculation was performed for a wavelength $\lambda = 0.1$m, target range $r_{tg} = 20$ km, antenna height $h_a = 10$m, and antenna elevation pattern width equal to $2°$ neglecting the diffraction [9]. The underlying surface was:

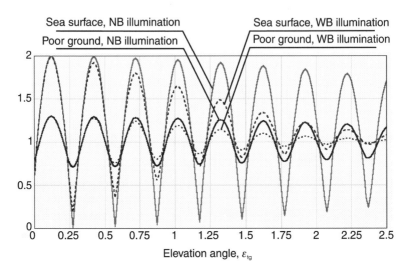

Figure 5.6 Pattern propagation factor calculated for narrowband (NB) illumination [1] and simulated for wideband (WB) illumination.

- Dry soil with a standard deviation in height equal to 0.1m covered by grass with $(1 - K_{Ab}) = 0.3$;
- Calm sea surface.

For a target elevation angle $\epsilon_{tg} < 0.5°$, results of calculating for the narrowband signal practically coincide with the results of simulation for the wideband signal. For a greater elevation angle ϵ_{tg}, the oscillations of the pattern propagation factor for the wideband signal are more damped than for the narrowband one. This can be explained by the increased relative shift of the envelopes of reflected and direct wideband signals, which are summed without mutual cancellation.

The oscillation frequency of the pattern-propagation factor $F(\epsilon_{tg})$ for various elevation angles ϵ_{tg} is determined by the rates of change of the path-delay difference $\delta t_l(\epsilon_{tg})$ between reflected and direct signals, their path-length difference $\delta r_l(\epsilon_{tg}) = c\delta t_l(\epsilon_{tg})$, and phase difference $2\pi f \delta t_l(\epsilon_{tg})$. The path-length differences $\delta r_l(\epsilon_{tg})$ and moduli of reflection coefficients versus elevation angle ϵ_{tg} are shown in Figure 5.7 for various underlying surfaces and polarizations. Pattern-propagation factor minimums are observed when the reflected and direct signals interfere out of phase. Their depth and also maximum amplitudes depend on the ratio of amplitudes of interfering signals. Evidently, these ratios take maximum values for a signal horizontal polarization and the surfaces with good conductivity. According

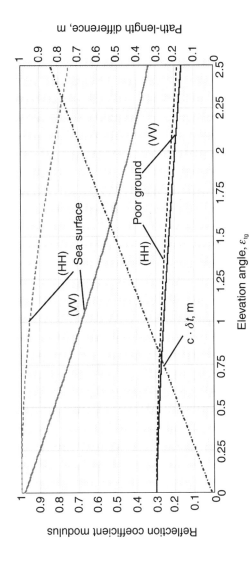

Figure 5.7 The path-length differences $\delta r_i(\epsilon_{tg})$ (dot-and-dash line) and reflection coefficient modulus for vertical (solid lines) and horizontal (dashed lines) polarizations and various underlying surfaces versus elevation angle ϵ_{tg}.

to (5.25), the modulus of reflection coefficient decreases with an increase of elevation angle (and the corresponding grazing angle).

Figure 5.8 shows the RPs of an F-15 aircraft [Figure 5.8(a), (c)] and of an ALCM missile [Figure 5.8(b), (d)] for the case of observation above the calm sea surface with the use of a 300-MHz chirp signal. Solid lines in Figure 5.8(a) and (b) correspond to undistorted RPs propagated through the radar-target-radar path, and dashed lines correspond to the delayed RPs reflected from the underlying surface. The path-length difference $\delta r_l(\epsilon_{tg})$ between these RPs for ϵ_{tg} = 2.5° was 0.856m. Results of interference of these signals accounting for their mutual phase differences are shown in Figure 5.8(c) and (d). The resultant RPs of the F-15 aircraft [Figure 5.8(c)] and the ALCM missile [Figure 5.8(d)] obtained after in-phase summation for the path-length difference of 0.856m are shown by solid lines, and those resultant RPs obtained after out-of-phase summation for the smaller path-length difference (about 0.1m) are shown by dotted lines. In the case of large path-length difference (relative to the range resolution element Δr), distortions of the RPs are determined mainly by the range shift of the signal envelopes. In the case of small path-length difference (comparable to the wavelength), the effect of the envelope differentiation exhibits itself, accompanied by significant energy losses.

The degree of the signal envelope distortion is characterized by the coefficient of correlation between the shapes of undistorted (free space) and distorted (due to surface reflections) RPs. Such correlation coefficients versus the target elevation angle ϵ_{tg} for the F-15 aircraft and the ALCM missile are shown in Figure 5.9, which also shows the pattern-propagation factor $F(\epsilon_{tg})$ versus the target elevation angle ϵ_{tg}. This figure justifies the conclusion that maximum distortion (decorrelation) of signals for small elevation angles of targets ϵ_{tg} (and small path-length differences) is caused by the effect of envelope differentiation when the direct and reflected signals interfere out of phase. For great ϵ_{tg} (and great path-length difference), the distortion due to range shift of envelopes becomes more significant. The RP correlation coefficient decreases to between 0.9 and 0.8 for well-conducting surfaces (calm sea surface, wet soil). Inversely, for poorly conducting surfaces (dry soil or other surfaces with diffuse principle scattering) distortions of the RP are not very significant.

5.3 Problem of the Wideband Target Recognition Under Conditions of Signal Distortions

The problem must be discussed under the assumption of the clutter cancellation. A wideband signal can be distorted by any kind of clutter canceller

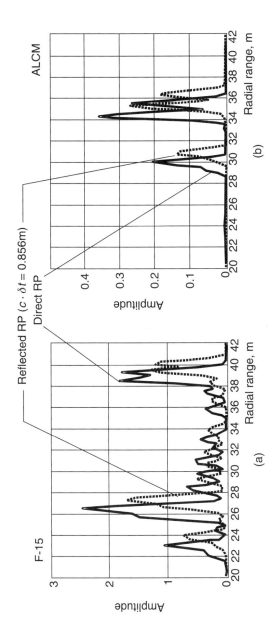

Figure 5.8 Simulated RPs for F-15 aircraft (a, c) and ALCM missile (b, d) observed above calm sea surface for various cases of propagation. In (a) and (b) solid lines represent the absence of reflection from the surface; dashed lines represent the presence of such reflection. In (c) and (d) solid lines represent in-phase summation of RPs (a, b); dashed lines represent out-of-phase summation of RPs (a, b).

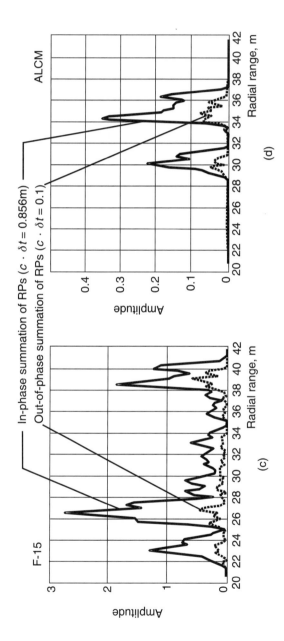

Figure 5.8 (continued).

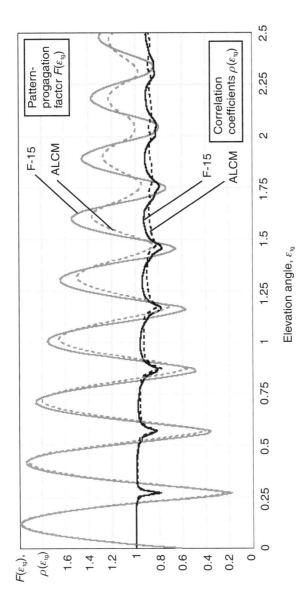

Figure 5.9 Correlation coefficients between undistorted (free space) and distorted (sea surface reflections) RPs, and the corresponding pattern propagation factors for two targets obtained using the 300-MHz chirp signal.

because of significant range shift of its envelope between two succeeding illuminations. This distortion can be much greater than corresponding distortions due to surface reflections (for dry soil or other surfaces with diffuse scattering) if the PRF is constant and not very high. The distortions of a clutter canceller can also be much smaller than the distortions due to surface reflections (wet soil or sea surface) if the PRF is high. We simulated, therefore, two cases of target recognition:

1. Target class recognition with distortions of RPs caused by the moving target indicator (MTI) only (Section 5.3.1);

2. Target type and class recognition with distortions of RPs caused by surface reflections only (Section 5.3.2).

The conditions of simulation were as follows: (1) range resolution of 0.5m, wavelength $\lambda = 0.1$m, and horizontal polarization; (2) yaw and pitch aspect were changed in the sectors of 10° to 30° and 2° to 10°, respectively; (3) target classes were recognized only by RPs using the correlation algorithm with the search within the range gate; (4) a dual MTI canceller was simulated; (5) rotational modulation was not introduced into the RPs; and (6) simulation was carried out assuming the recognition provided by a single illumination under conditions with an absence of noise and clutter residue. Let us mention that in Chapter 4 some examples were given showing that the influence of the time varying realizations of noise on recognition can be substantially reduced by an increase in the transmitted energy and the number of illuminations and recognition signatures.

5.3.1   Target Class Recognition for the RP Distortions by MTI Only

Wideband recognition by use of RPs was simulated for a B-52 and an F-15 aircraft and for an ALCM missile. Recognition of classes was reduced in this case to recognition of target types provided there was only one type of target in each class. The objective of the study was the dependence of recognition quality on the target radial velocity using a dual-canceller MTI with a PRF of 365 Hz.

The MTI influence on the RP structure is as follows. The RPs at the input (solid line) and output (bars) of dual-canceller MTI are shown in Figure 5.10 for large-sized (B-52), medium-sized (F-15), and small-sized (ALCM) targets. Radial velocities V_r for simplicity were chosen to be the same for all the targets and equal to 300 m/s, so that the ratio

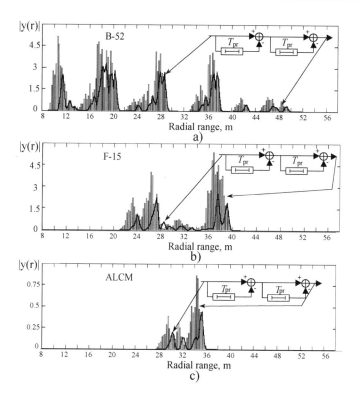

Figure 5.10 Range profiles at the input (lines) and output (bars) of MTI for (a) B-52 and (b) F-15 aircraft and for (c) ALCM missile.

$(V_r T_{pr})/\Delta r \approx 1.6$. It can be seen from Figure 5.10 that the range profiles are stretched and distorted at the MTI output.

Amplitude-velocity characteristics of the dual-canceller simulated for large-sized and small-sized targets using wideband illumination signals are shown in Figure 5.11.

It can be seen from Figure 5.11 that the intervals between the maximum and minimum values of response of the dual-canceller MTI and its mean value decrease as the target radial velocity increases.

Recognition using the RPs distorted by MTI was simulated using three individualized standard RPs for each target obtained by the sets of 100 teaching RPs. Reference conditional probabilities of class recognition for single ($N = 1$) target illumination using the undistorted RPs are presented in Table 5.4. Corresponding conditional probabilities of class recognition using the RPs distorted by MTI, but without accounting for the influence of such distortion, are shown in Table 5.5 for the ratio $(V_r T_{pr})/\Delta r \approx 1.6$. It is seen that recognition quality decreases.

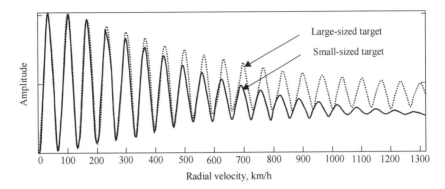

Figure 5.11 Amplitude-velocity characteristics of dual-canceller simulated for large-sized and small-sized targets using wideband illumination signal.

Table 5.4
Reference Conditional Probabilities of Target Recognition for Single (*N* = 1) Target Illumination Using the RPs and Standards Undistorted in the MTI

	Solution		
Condition	B-52	F-15	ALCM
B-52	0.98	0.02	0.0
F-15	0.0	1.0	0.0
ALCM	0.0	0.0	1.0

Table 5.5
Conditional Probabilities of Target Recognition for Single (*N* = 1) Target Illumination Using the RPs Distorted in the MTI and Undistorted Standards

	Solution		
Condition	B-52	F-15	ALCM
B-52	0.84	0.09	0.07
F-15	0.11	0.75	0.14
ALCM	0.0	0.04	0.96

To avoid such decreases in recognition quality, one can anticipate the preliminary distortion of the RPs introduced by the MTI under assumptions of various target velocities, and then increase the number of standards according to the number of velocity channels. At the recognition stage, one has to obtain a rough velocity estimate $(\sigma_{V_r} T_{pr} \ll \Delta r)$ together with a coarse estimate of the target aspect sector.

Table 5.6 presents the conditional probabilities of recognition corresponding to the use of the sets of teaching RPs preliminarily distorted by the MTI under the assumption of a target velocity interval of 300 ± 50 m/s and $(2m\Delta V_r T_{pr})/\Delta r < 0.5$, where m is the number of the canceller delays. The quality of recognition is increased compared to Table 5.5.

5.3.2 Target Type and Class Recognition for the RP Distortions by Underlying Surface Only

Recognition of 11 Target Types. Wideband target recognition by RPs was simulated for the following targets: Tu-16, B-52, B-1B, Mig-21, F-15, Tornado, and An-26 aircraft; AH-64 helicopter; ALCM and GLCM missiles; and passive decoy.

Reference conditional probabilities of type recognition without any distortions (free space, no noise) are presented in Table 5.7. The results presented correspond to the use of three standard RPs for each type of target obtained by the sets of 100 teaching RPs. The whole probability of error was $P_{er} = 0.03$. Table 5.8 presents corresponding conditional probabilities of recognition for the case where the RPs were distorted by the reflections from the underlying surface and standard RPs were assumed to be the same as for free space. Targets were observed above a calm sea surface at an elevation angle of $\epsilon_{tg} = 2.5°$. The probability of error in type recognition increased to $P_{er} = 0.275$. Such a great increase was due to an increased number of errors in recognition of small-sized targets, RPs of which were stretched so that they were misrecognized as medium-sized targets.

Recognition of Three Target Classes. Analogous results were obtained for recognition of the large-sized, medium-sized, and small-sized target classes

Table 5.6
Conditional Probabilities of Target Recognition for Single ($N = 1$) Target Illumination Using the RPs and Standards Distorted in the MTI

	Solution		
Condition	**B-52**	**F-15**	**ALCM**
B-52	0.97	0.03	0.0
F-15	0.01	0.97	0.02
ALCM	0.0	0.0	1.0

Table 5.7
Conditional Probabilities of the 11 Target Type Recognitions for Single ($N = 1$) Target Illumination by Undistorted RPs
(Free Space, No Noise)

Condition	Decision										
	Tu-16	B-52	B-1B	MiG-21	F-15	Tornado	ALCM	GLCM	An-26	Decoy	AH-64
Tu-16	**0.96**	0.03	0.01	0	0	0	0	0	0	0	0
B-52	0	**0.95**	0.05	0	0	0	0	0	0	0	0
B-1B	0	0.06	**0.88**	0.04	0	0	0	0	0	0.02	0
MiG-21	0	0	0	**0.98**	0.02	0	0	0	0	0	0
F-15	0	0	0	0.02	**0.97**	0.01	0	0	0	0	0
Tornado	0	0	0	0.04	0	**0.95**	0.01	0	0	0	0
ALCM	0	0	0	0	0	0	**1**	0	0	0	0
GLCM	0	0	0	0	0.02	0	0.02	**0.87**	0	0.09	0
An-26	0	0	0	0	0	0	0	0	**0.79**	0.17	0.04
Decoy	0	0	0	0	0	0	0	0	0	**1**	0
AH-64	0	0	0	0	0	0	0	0.01	0.11	0.02	**0.86**

Table 5.8
Conditional Probabilities of the 11 Target Type Recognitions for Single ($N = 1$) Target Illumination by Distorted RPs
(Reflections from Sea Surface, No Noise)

Condition	Tu-16	B-52	B-1B	MiG-21	F-15	Tornado	ALCM	GLCM	An-26	Decoy	AH-64
						Decision					
Tu-16	**0.91**	0.03	0.01	0.01	0.04	0	0	0	0	0	0
B-52	0.11	**0.67**	0.06	0.12	0.03	0.01	0	0	0	0	0
B-1B	0.03	0.09	**0.8**	0.06	0.02	0	0	0	0	0	0
MiG-21	0.01	0	0	**0.9**	0.04	0.05	0	0	0	0	0
F-15	0	0	0	0.06	**0.87**	0.07	0	0	0	0	0
Tornado	0.01	0	0	0.05	0.03	**0.91**	0.01	0	0	0	0
ALCM	0	0	0	0	0.08	0	**0.02**	0.26	0.64	0	0
GLCM	0	0	0	0	0.0	0	0.07	**0.82**	0.01	0.1	0
An-26	0	0	0	0	0.08	0	0.02	0.22	**0.63**	0	0.05
Decoy	0	0	0	0	0	0	0.17	0.83	0	**0**	0
AH-64	0	0	0	0	0	0	0.07	0.17	0.14	0	**0.62**

by undistorted (Table 5.9) and distorted (Table 5.10) RPs. It is seen that the increase in probability of error from P_{er} = 0.01 to P_{er} = 0.23 is due to conditional probability of recognition for small-sized targets being reduced from 0.99 to 0.42.

Conclusions. The results of Tables 5.8 and 5.10 were obtained for the worst case where the targets were observed above a calm sea surface using large signal bandwidth of 300 MHz. The decrease of recognition quality observed in this case relative to free space can probably be recovered by decreasing the signal bandwidth and by using preliminary distortions of standard RPs (similar to the case of MTI, Section 5.3.1). For many aspects of shortwave illumination, the medium-sized targets can be distinguished from the small ones using a rotational modulation signature. Further simulations can help, apparently, in developing methods for low-altitude target recognition over the sea.

For the targets observed above a dry surface, the simulation carried out shows that the correlation coefficient of RPs is not significantly decreased,

Table 5.9
Conditional Probabilities of the Three Target Class Recognitions for Single (N = 1) Target Illumination by Undistorted RPs (Free Space, No Noise)

	Solution		
Condition	**Large-sized**	**Medium-sized**	**Small-sized**
Large-sized	**0.987**	0.013	0
Medium-sized	0	**0.997**	0.003
Small-sized	0	0.007	**0.993**

Table 5.10
Conditional Probabilities of the Three Target Class Recognitions for Single (N = 1) Target Illumination by Distorted RPs (Reflections from Sea Surface, No Noise)

	Solution		
Condition	**Large-sized**	**Medium-sized**	**Small-sized**
Large-sized	**0.907**	0.093	0
Medium-sized	0.007	**0.99**	0.003
Small-sized	0	0.58	**0.42**

and the recognition quality is not significantly reduced compared to the case of recognition in free space (Table 5.7).

References

[1]   Barton, D. K., *Modern Radar System Analysis*, Norwood, MA: Artech House, 1988.

[2]   Cherniy, F. B., *Propagation of Radio Waves*, Moscow: Sovetskoe Radio Publishing House, 1972 (in Russian).

[3]   Kulemin, G. P., and V. B. Razskazovsky, *Millimeter Radio Waves Scattering on Earth Ground Under Low Angles*, Kiev: Naukova Dumka Publishing House, 1987 (in Russian).

[4]   Kulemin G. P., "Radar Clutter from Sea and Ground on Centimeter and Millimeter Waves," *Proceedings of the "Modern Radar" Int. Conference*, Kiev, Russia, November 1994, Part 1, pp. 27–32.

[5]   http://info.er.usgs.gov/reseach/gis/title.html, retrieved January 12, 2000.

[6]   http://service.uga.edu.narsal/gis.html, retrieved January 12, 2000.

[7]   Smirnov, V. A., *Dynamics of Motion of Wheel-Driven Vehicles*, Moscow: Mashinostroenie Publishing House, 1989 (in Russian).

[8]   Slavutsky, A. K., *Design, Construction, Maintenance and Repair of Agricultural Roads*, Moscow: Vysshaya Shkola Publishing House, 1972 (in Russian).

[9]   Fock, V. A, *Problems of Diffraction and Propagation of Electromagnetic Waves*, Moscow: Sovetskoe Radio Publishing House, 1970 (in Russian).

[10]   Kalinin, A. I., and E. L. Tcherenkova, *Propagation of Radio Waves and Operation of Wireless Lines*, Moscow: Svyaz Publishing House, 1971 (in Russian).

[11]   Bakhvalov, B.N., *Reference Materials to Evaluating the Influence of Real Positions of Radar Systems on Their Coverage Zones*, Kharkov: Military Radio Engineering Academy, 1977 (in Russian).

[12]   Kerr, D. E. (ed), *Propagation of Short Radio Waves*, M.I.T. Radiation Laboratory Series, No. 13, New York: McGraw-Hill, 1951.

[13]   Bass, F. G., and I. M. Fuks, *Diffraction of Waves from Statistically Nonuniform Surfaces*, Moscow: Nauka Publishing House, 1972 (in Russian).

6

Review and Simulation of Signal Detection and Operation of Simplest Algorithms of Target Tracking

Since the basic algorithms of detection and tracking are well known, we will deal mainly with the discussion of the possible contribution of backscattering simulation methods in the Research and Development (R&D) of corresponding radar systems. In Section 6.1 we compare an a priori and simulated fluctuation pdf of a narrowband radar echo signal and consider its detection characteristics based on this pdf [1–13]. In Section 6.2 we consider target coordinate and doppler glints for narrowband illumination and their influence on tracking with corresponding results of simulation [14–19]. In Section 6.3 we discuss some aspects of wideband signal use in detection and tracking based on simulation results [20, 21]. The "log-scale" method and some other methods for the detection of various targets illuminated by wideband signals are proposed and their quality indices are evaluated.

6.1 Target RCS Fluctuations and Signal Detection with Narrowband Illumination

In this section we consider the background, details, and statement of the problem (Section 6.1.1), list the variants of simulation of signal detection on the noise background (Section 6.1.2), and compare, using examples, the simulated pdf of RCS with its a priori pdf (Section 6.1.3).

6.1.1 Background, Details, and Statement of the Problem

Initial Swerling Distributions. Both the radar designer and radar analyst need the target RCS to be properly specified for evaluation of radar performances in various conditions. The detection range of a fluctuating target in a noise background is one of such valuable performance measures. The problem of its evaluation was systematically studied after World War II, but generalized results of this study were published only in 1960 [1, 2].

For the coherent signals, Swerling considered two types of pdf of the ratio $x = \sigma/\overline{\sigma} > 0$, where σ is the target RCS and $\overline{\sigma}$ is its mean value [2]. These pdfs were

$$p(x) = e^{-x} \quad \text{and} \quad p(x) = 4xe^{-2x} \tag{6.1}$$

The value of both pdfs (6.1) resulted from their adequate approximation of real situations for definite target classes. The first pdf of (6.1) describes the RCS fluctuations of a large-sized target in nonradial flight, when the variance of RCS is great. The second pdf of (6.1) describes the RCS fluctuations for a medium-sized target also in nonradial flight.

Chi-Square pdf. In limited aspect sectors, the variance of RCS decreases. Therefore, in 1956–1957 Swerling and Weinstock [3, 4] proposed to use a more general chi-square pdf of ratio $x = \sigma/\overline{\sigma} > 0$ with $2k$ degrees of freedom:

$$p_k(x) = A_k x^{k-1} e^{-kx} \tag{6.2}$$

where $A_k = k^k/(k - 1)!$ This corresponds to a pdf of the normalized signal amplitude b:

$$p_k(b) = 2A_k b^{2k-1} e^{-kb^2} \quad E(b^2) = \overline{b^2} = 1 \tag{6.3}$$

Both previous pdfs (6.1) became special cases of pdf (6.2). The first pdf of (6.1) corresponds to the value $k = 1$ in (6.2) and (6.3). The amplitude distribution (6.3) for $k = 1$ is the well-known Rayleigh distribution. The second pdf of (6.1) corresponds to the value $k = 2$ in (6.2) and (6.3). Detection probabilities for various k in (6.2) and (6.3) were presented in tables of Meyer and Mayer's handbook [5]. For large k both pdfs, (6.2) and (6.3), approach the Gaussian.

Log-Normal pdf. Heidbreder and Mitchell [6] showed in 1967 that the chi-square model (6.2) is inadequate for description of a significantly asymmetric

RCS pdf such as that of a missile or a ship. As a measure of asymmetry, they introduced the ratio

$$R = \overline{\sigma}/\sigma_{\text{med}}$$

of the mean and median values of RCS. The median value σ_{med} is determined so that the probabilities of the random RCS values σ exceeding σ_{med} and of those being below σ_{med} are identical and equal to 0.5. The ratio R is equal to unity for the Gaussian pdf, equal to 1.18 for the second of pdf (6.1), and equal to 1.44 for the first of pdf (6.1). However, the ratio R for real missiles and ships is much greater. As the pdf applicable for various values of R, Heidbreder and Mitchell proposed to use the log-normal pdf $p(x)$, $x = \sigma/\overline{\sigma} > 0$ obtained on the basis of a Gaussian pdf for the value $y = \ln x$, so that

$$p(x) = p(y)\left|\frac{dy}{dx}\right| = \frac{1}{\sqrt{2\pi D}} \exp\left[-\frac{(\ln x - M)^2}{2D}\right]\left|\frac{d\ln x}{dx}\right|$$

or

$$p(x) = \frac{1}{x\sqrt{2\pi D}} \exp\left[-\frac{(\ln x - M)^2}{2D}\right] \tag{6.4}$$

where the parameters M and D are the mean and variance of pdf of the random value $y = \ln x$. Let us find them as functions of ratio R.

It is known that for the Gaussian pdf of y, its mean value (mathematical expectation) $\overline{y} = M$ coincides with its median value y_{med} or $M = y_{\text{med}}$. In turn, owing to the monotonic nature of the logarithmic function $y = \ln x$, the median value y_{med} coincides with the value of function $y_{\text{med}} = \ln(x_{\text{med}})$ of median value of argument x_{med}, so $M = \ln(x_{\text{med}})$. Since the median value of the variable $x = \sigma/\overline{\sigma}$ is $x_{\text{med}} = \sigma_{\text{med}}/\overline{\sigma}$, then

$$M = \ln(\sigma_{\text{med}}/\overline{\sigma}) = \ln R^{-1} = -\ln R$$

The mean value of $x = \sigma/\overline{\sigma}$ $(0 \le x < \infty)$ is $\overline{x} = \overline{\sigma}/\overline{\sigma} = 1$. But in integral form it is

$$\overline{x} = \int_0^\infty xp(x)dx = 1$$

The latter equation allows expressing the unknown parameter D of pdf (6.4) through the ratio $R = \overline{\sigma}/\sigma_{med}$. Using (6.4) and replacing the variable x with a new variable z by means of the relation $x = R^{-1}e^{\sqrt{D}(z+\sqrt{D})}$, $-\infty < z < \infty$, we obtain

$$\frac{1}{\sqrt{2\pi D}}\int_0^\infty \exp\left\{-\frac{[\ln(Rx)]^2}{2D}\right\}dx = R^{-1}e^{D/2} \cdot \frac{1}{\sqrt{2\pi}}\int_{-\infty}^\infty e^{-z^2/2}dz = R^{-1}e^{D/2} = 1$$

or $D = 2\ln R$.

So, having determined $R = \overline{\sigma}/\sigma_{med}$ by the experiment or simulation, one can define the parameters of log-normal pdf (6.4) as

$$M = -\ln R \text{ and } D = 2\ln R.$$

Other Propositions, Statement of the Problem. In 1997 a new discussion of the old RCS pdf problem took place. It was initiated by Johnston [8] and participated in by Swerling [9].

Xu and Huang [10] proposed to approximate the RCS pdf by the sum of 10 to 30 Legendre orthogonal polynomials with coefficients depending sophistically on the central moments of the real target pdf.

Johnston [11] intended to simulate the target backscattering. The following requirements to a desirable model were formulated: "The target model should consist of three submodels: RCS, motion and glint. These must be associated with an environmental model; this association must be done for each specific type or family of targets."

By then, the authors of this book had been working for many years in formulating such models for simulation, especially in recognition. The results of this work could be used in detection and tracking, as was stated in [12, 13] and will be considered below.

6.1.2 Variants of Simulation of Signal Detection on the Noise Background

There are several variants of detection probability evaluation:

- Calculation of detection probabilities in a noise background using the simulated RCS pdf for theoretical signal models. For this case we simulate the RCS pdf (Section 6.1.3);

• Complex simulation of the target backscattering in test flights at given altitudes. Corresponding examples for detection of wideband signals in free space will be considered in Section 6.3.1. Simulation of detection of narrowband and wideband signals can be provided also in conditions of propagation near the Earth's surface (Chapter 5) for various cases of MTI and MTD use.

6.1.3   The Simulated RCS pdf and Comparison with Its A Priori pdf

The simulated pdf of RCS depends on the type of target and on the given aspect sector. Simulated RCS histograms obtained in the 0° to 20° aspect sector are shown in Figure 6.1 for Tu-16-type aircraft and ALCM-type missile. Histogram approximations are also shown in Figure 6.1. The chi-square pdf approximates the histogram in Figure 6.1(a) for the aircraft, and the log-normal pdf approximates the histogram in Figure 6.1(b) for the missile. The number of duo-degrees of freedom for the Tu-16 bomber in Figure 6.1(a) is $2K = 1.6$ (the same as for experimental data for "typical aircraft" of unspecified type [10]).

The simulated RCS histogram approximations obtained in the 0° to 20°, 30° to 50°, and 60° to 80° aspect sectors for the Tu-16- and F-15-type aircraft and the ALCM-type missile are shown in Figure 6.2. The histograms were approximated with a chi-square pdf for Tu-16- and F-15-type aircraft

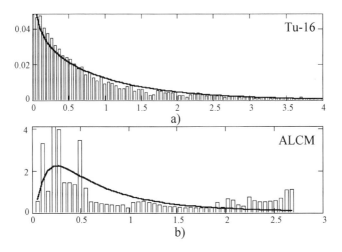

Figure 6.1   The RCS distribution histograms and their approximations: (a) for Tu-16-type aircraft approximated by chi-square pdf; and (b) for ALCM missile approximated by log-normal pdf.

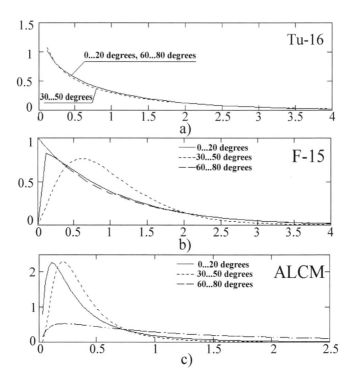

Figure 6.2 Approximations of the RCS histograms for various aspect sectors: (a) with chi-square pdf for the Tu-16-type aircraft; (b) with chi-square pdf for the F-15-type aircraft; and (c) with log-normal pdf for the ALCM-type missile.

and with a log-normal pdf for ALCM-type missile. Influence of the aspect sector choice on the RCS pdf is weak for the large-sized aircraft (Tu-16 type). Conversely, it is strong for the medium-sized aircraft (F-15 type) and small-sized missile (ALCM type).

The detection simulation shows:

1. Closeness of the simulation results to the experimental data of other authors;

2. Advantage of backscattering models taking into account the specific features of the target type and its aspect sector in comparison with the widely used pdf approximations.

6.2 Coordinate and Doppler Glint in the Narrowband Illumination

R&D of tracking systems consider various error sources: thermal noise, multipath, irregularity of atmosphere refraction, and the glint of extended targets in angle, range (delay), and doppler frequency [14]. The glint sources can easily be considered with the backscattering model used with or without accounting for thermal noise. Extended targets glint can be simulated for both sequential scan type and monopulse type angle measurement, as well as for single or integrated measurement results in view of target motion. In Section 6.2.1 we consider the extended target concept and basic glint equations for narrowband signals in free space. In Section 6.2.2 we consider examples of the theoretical glint analysis for the two-element target model. Angular, delay, doppler frequency glints, and the concept of a center of the target backscattering will be considered particularly in Sections 6.2.1 and 6.2.2. Possible simulation results for the glint of complex targets will be given and discussed in Sections 6.2.3 and 6.2.4. Using Chapter 5, one can consider also additional glint under condition of reflections from the underlying surface.

6.2.1. The Extended Target Concept and Basic Equations of Target Glint

The extended target concept depends on the kind of illumination. With narrowband illumination, the "glint" can be especially intense. The inherent random components of errors in measurement of coordinates and velocities caused by the interference of reflections from different elements of the complex target are so named.

A target with unresolved elements is considered as extended if its glint exceeds or approaches the equipment errors of the coordinate measurement. Angular, range, spatial, and doppler glint can be taken into account [14–19]. Let us consider all of them, as usual, without accounting for rotational modulation, which will be accounted for separately in simulation.

Angular Glint. Angular glint can be considered as the change of phase front orientation of the backscattered wave in the vicinity of the receiving antenna due to the interference phenomenon. The surface of constant phase

$$\chi(\mathbf{L}, f, t) = \arg \dot{E}(\mathbf{L}, f, t) = \text{const}$$

is assumed here to be the phase front for narrowband illumination, and $\dot{E}(\mathbf{L}, f, t)$ (Section 1.3.4) is considered as a result of wave interference for a bistatic radar. This result is defined by (1.20), where one can assume $U(t - \Delta t) \approx U(t)$ for narrowband illumination. We assume here that the target elements are not resolved in angle since antenna dimensions are limited, and hence the phase front of the received wave can be regarded as flat in the antenna vicinity. The unit vector $\mathbf{U}^0$ of the partial phase gradient, being orthogonal to this phase front, defines the direction of wave arrival and characterizes the apparent direction of the target. The unit vector $\mathbf{U}^0$, plotted from the receiving position of a bistatic radar "in the target direction" [Figure 6.3(a)], is defined by

$$\mathbf{U}^0 = \mathbf{U}^0(\mathbf{L}, f, t) = \frac{\partial \chi(\mathbf{L}, f, t)}{\partial \mathbf{L}} \bigg/ \left| \frac{\partial \chi(\mathbf{L}, f, t)}{\partial \mathbf{L}} \right| \tag{6.5}$$

Equation (6.5) will be used mainly in the case of monostatic radar, where $\mathbf{U}^0(\mathbf{L}, f, t)$ calculated for $\mathbf{L} \to 0$ shows the target apparent direction. The difference vector $\mathbf{U}^0(0, f, t) - \mathbf{R}^0(t)$ describes the absolute value and the components of angular glint in azimuthal and elevation planes. The absolute value $\Delta\theta(t)$ of small angular glint [Figure 6.3(b)] is equal approximately to the absolute value of this difference vector

$$\Delta\theta(t) \approx \left| \mathbf{U}^0(0, f, t) - \mathbf{R}^0(t) \right| \tag{6.6}$$

The azimuth glint $\Delta\beta(t)$ and the elevation glint $\Delta\epsilon(t)$ can be found approximately as the components of vector $\mathbf{U}^0(0, f, t) - \mathbf{R}^0(t)$.

Group Delay. The group delay concept permits us to calculate and clarify the range glint for narrowband signals. The transmitter-target-receiver propa-

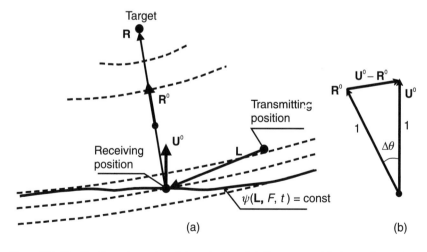

Figure 6.3 Clarification of angular glint evaluation: (a) disagreement between the vectors $\mathbf{U}^0$ and $\mathbf{R}^0$ of apparent and actual directions; and (b) formation of angular glint.

gation path can be considered as a linear four-terminal network. For a very narrowband signal, its phase-frequency response (Figure 6.4) is linear

$$\arg K(f + F) = \arg K(f) + Fd[\arg K(f)]/df \qquad (6.7)$$

and its amplitude-frequency response is constant.

According to the rules of trigonometry, each sinusoidal component of a narrowband signal can be presented as $A_k \cos[2\pi(f + F_k)t + \chi_k] = A_k \cos(2\pi ft) \cos(2\pi F_k t + \chi_k) - A_k \sin(2\pi ft) \sin(2\pi F_k t + \chi_k)$. Superposition of such components $k = 1, 2, \ldots$ can be presented in the form

$$A(t)\cos(2\pi ft) - B(t)\sin(2\pi ft) = C(t)\cos[2\pi ft + \chi(t)]$$

where

$$A(t) = \sum_k A_k \cos(2\pi F_k t + \chi_k), \quad B(t) = \sum_k A_k \sin(2\pi F_k t + \chi_k) \qquad (6.8)$$

$$C(t) = \sqrt{A^2(t) + B^2(t)}, \qquad \chi(t) = \operatorname{atan}[B(t)/A(t)]$$

Each of the input signal components $A_k \cos[2\pi(f + F_k)t + \chi_k]$ at the output of a linear four-terminal network, after obtaining phase delays (6.7), takes the appearance

$$A_k \cos\left\{2\pi ft - \arg K(f) + \chi_k + 2\pi F_k \left[t - \frac{1}{2\pi}\frac{d\arg K(f)}{df}\right]\right\}$$

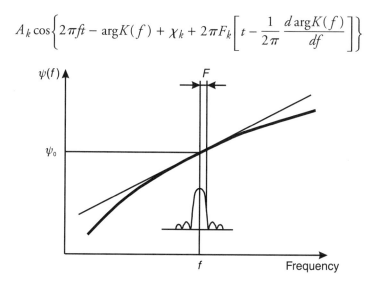

Figure 6.4 Clarification of the conditions for introducing the group delay concept.

Superposition of the output components can be presented in the form

$$A(t - t_{gr}) \cos[2\pi ft - \arg K(f)] - B(t - t_{gr}) \sin[2\pi ft - \arg K(f)]$$
$$= C(t - t_{gr}) \cos[2\pi ft + \chi(t - t_{gr}) - \arg K(f)] \qquad (6.9)$$

Here, $A(t)$, $B(t)$, $C(t)$, and $\chi(t)$ are the time functions defined by (6.8).

The t_{gr}, named the group delay, is a delay of the signal complex envelope $C(t) \exp[j\chi(t)]$. It is proportional to the derivative of the argument $\arg K(f)$ of complex amplitude-frequency response of the propagation path

$$t_{gr} = \frac{1}{2\pi} \frac{d}{df} \arg K(f) \qquad (6.10)$$

It is significant that the complex envelope (6.9) of the delayed signal is not distorted with narrowband illumination. Inversely, the resolution of target elements arising with wideband illumination can be considered here as a specific kind of distortion of the input signal.

Range Glint. Range glint in narrowband illumination is a simple consequence of the corresponding group delay glint Δt_{gr} caused by the interference effects of backscattering. Range glint $\Delta R(t)$ is defined by

$$\Delta R(t) = \frac{c}{2} \Delta t_{gr}(t) = \frac{c}{4\pi} \frac{\partial}{\partial f} \arg \dot{E}(\mathbf{L}, f, t) - |\mathbf{R}(t)| \qquad (6.11)$$

The minuend of the right-hand part of (6.11) describes the apparent target range $R_a(t)$; $\mathbf{R}(t)$ is the radius vector plotted from the origin of the radar coordinate system to the origin of the target system, which is assumed to be near to the target geometrical center.

Spatial Glint. Spatial glint with narrowband illumination is defined approximately as a sum of vectors

$$\mathbf{a}(t) \approx \frac{c}{2} \Delta t_{gr}(t) \mathbf{U}^0(0, f, t) + R_a(t)[\mathbf{U}^0(0, f, t) - \mathbf{R}^0(t)] \qquad (6.12)$$

The first term accounts for the glint (6.11) in the apparent radial direction, and the second term accounts for the glint (6.12) in the apparent transverse direction. Each of two apparent displacements and the whole one $|\mathbf{a}(t)|$ often exceed the target dimensions.

Doppler Glint. Doppler glint is caused by translational and rotational target motion. It is defined by the partial time derivative of the backscattered signal argument

$$\Delta F_D = \frac{1}{2\pi} \frac{\partial \chi(0, f, t)}{\partial t} - F_{D0}(t) \qquad (6.13)$$

The minuend of the right-hand part of (6.13) describes the apparent doppler frequency $F_D(t)$. The subtrahend $F_{D0}(t)$ is the doppler frequency of a point backscatterer having no doppler glint.

6.2.2 Examples of the Theoretical Analysis of Glint for Two-Element Target Model

The target model (Figure 6.5) observed with a monostatic radar consists of two unresolved point elements backscattering independently with RCS $\sigma_{1,2}$, respectively. Positions of the point elements of the model relative to the origin O_{tg} of the local coordinate system are described by the relatively small vectors $\mp\mathbf{d}/2$. We also will use the bistatic radar coordinate system, whose origin O_{rad} coincides with the transmitting position. Receiving position is described in this system by vector $\mathbf{L}$, where for monostatic radar $\mathbf{L} \to 0$. The origin O_{tg} of target coordinate system is described in the radar one by the vector $\mathbf{R} = \mathbf{R}(t) = R(t)\mathbf{R}^0(t)$, where $\mathbf{R}^0(t)$ is the unit vector.

Angular Glint Evaluation. This will be carried out using (6.5). The equation includes the phase dependence in a bistatic radar $\chi(\mathbf{L}, f, t)$, where $\mathbf{L}$ is the

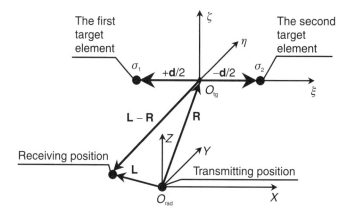

Figure 6.5 Geometry for evaluation of angular and spatial glint for the two-element model of the target.

vector plotted between transmitting and receiving positions. In the case of monostatic radar, we can assume that $\mathbf{L} \to 0$, but only in the final equations. According to (1.20), a sinusoidal signal $\dot{E}(\mathbf{L}, f, t)$ backscattered from the target model (up to a constant multiplier) can be presented in trigonometrical form, which can be transformed by analogy with (6.8):

$$
\begin{aligned}
E(\mathbf{L}, f, t) &= \mathrm{Re}[\dot{E}(\mathbf{L}, f, t)\exp(j2\pi ft)] \\
&= \sqrt{\sigma_1}\cos[2\pi f(t - \Delta t_1)] + \sqrt{\sigma_2}\cos[2\pi f(t - \Delta t_2)] \quad (6.14) \\
&= A(\mathbf{L})\cos(2\pi ft) + B(\mathbf{L})\sin(2\pi ft)
\end{aligned}
$$

Here, $A(\mathbf{L}) = \sqrt{\sigma_1}\cos(2\pi f\Delta t_1) + \sqrt{\sigma_2}\cos(2\pi f\Delta t_2)$ and $B(\mathbf{L}) = \sqrt{\sigma_1}\sin(2\pi f\Delta t_1) + \sqrt{\sigma_2}\sin(2\pi f\Delta t_2)$ are the amplitudes of total quadrature components of the signal; and $\Delta t_{1,2}$ are the delays of partial signals backscattered by the first and second target model elements displaced from the target model center by $\mp \mathbf{d}/2$. Each of them is composed of the delay on the direct path $|\mathbf{R} \mp \mathbf{d}/2|$ and the delay on the back path $|\mathbf{R} - \mathbf{L} \mp \mathbf{d}/2|$. So,

$$
\Delta t_{1,2}(\mathbf{L}) = [|\mathbf{R} \mp \mathbf{d}/2| + |\mathbf{R} - \mathbf{L} \mp \mathbf{d}/2|]/c \quad (6.15)
$$

Since $|\mathbf{L}| \ll |\mathbf{R} \mp \mathbf{d}/2|$, let us use an approximation

$$
|\mathbf{S} - \Delta\mathbf{S}| \approx |\mathbf{S}| - \Delta\mathbf{S}^{\mathrm{T}}\mathbf{S}^0 \quad (6.16)
$$

where $\mathbf{S}^0 = \mathbf{S}/|\mathbf{S}|$ is a unit vector. Approximation (6.16) is correct for an arbitrary vector $\Delta\mathbf{S}$ with relatively small modulus $|\Delta\mathbf{S}| \ll |\mathbf{S}|$ and can be easily affirmed geometrically or in coordinate form. Using the approximation (6.16), we obtain

$$
\Delta t_{1,2}(\mathbf{L}) \approx [2|\mathbf{R} \mp \mathbf{d}/2| - \mathbf{L}^{\mathrm{T}}(\mathbf{R} \mp \mathbf{d}/2)^0]/c \quad (6.17)
$$

The phase distribution of a backscattered wave in the vicinity of a monostatic radar $\mathbf{L} \to 0$ is defined by the function $\chi(\mathbf{L}, f, t) = \mathrm{atan}[B(\mathbf{L})/A(\mathbf{L})]$. Its partial gradient

$$
\begin{aligned}
\left.\frac{\partial\chi(\mathbf{L}, f, t)}{\partial\mathbf{L}}\right|_{\mathbf{L}=0} &= \left.\frac{\partial}{\partial\mathbf{L}}\mathrm{atan}\frac{B(\mathbf{L})}{A(\mathbf{L})}\right|_{\mathbf{L}=0} \\
&= \frac{A(0)[dB(0)/d\mathbf{L}] - B(0)[dA(0)/d\mathbf{L}]}{A^2(0) + B^2(0)}
\end{aligned} \quad (6.18)
$$

defines the apparent direction of wave arrival. To calculate the derivatives contained in (6.18), let us find first the derivatives (gradients) of vector $\mathbf{L}$ of (6.17) under the condition of $|\mathbf{d}| \ll |\mathbf{R}|$. They are

$$\frac{\partial \Delta t_{1,2}(0)}{\partial \mathbf{L}} = -\frac{\mathbf{R} \mp \mathbf{d}/2}{c|\mathbf{R} \mp \mathbf{d}/2|} \approx -\frac{\mathbf{R} \mp \mathbf{d}/2}{c(|\mathbf{R}| \mp \mathbf{d}^T\mathbf{R}^0/2)} \tag{6.19}$$

$$\approx \frac{1}{c}\left[\mathbf{R}^0\left(1 \pm \frac{\mathbf{d}^T\mathbf{R}^0}{2|\mathbf{R}|}\right) \mp \frac{\mathbf{d}}{2|\mathbf{R}|}\right]$$

Using (6.19), we find the necessary derivatives (gradients)

$$\frac{dB(0)}{d\mathbf{L}} = \frac{2\pi}{\lambda}\left\{\mathbf{R}^0\left[A(0) + \frac{\mathbf{d}^T\mathbf{R}^0}{2|\mathbf{R}|}A_1(0)\right] - \frac{\mathbf{d}}{2|\mathbf{R}|}A_1(0)\right\} \tag{6.20}$$

$$\frac{dA(0)}{d\mathbf{L}} = -\frac{2\pi}{\lambda}\left\{\mathbf{R}^0\left[B(0) + \frac{\mathbf{d}^T\mathbf{R}^0}{2|\mathbf{R}|}B_1(0)\right] - \frac{\mathbf{d}}{2|\mathbf{R}|}B_1(0)\right\}$$

where $A_1(0) = \sqrt{\sigma_1}\cos(2\pi f\Delta t_1) - \sqrt{\sigma_2}\cos(2\pi f\Delta t_2)$ and $B_1(0) = \sqrt{\sigma_1}\sin(2\pi f\Delta t_1) - \sqrt{\sigma_2}\sin(2\pi f\Delta t_2)$. Substituting the expressions (6.20) with given values of $A(0)$, $B(0)$, $A_1(0)$, and $B_1(0)$ into (6.18), we obtain after trigonometric transformations of the partial gradient of function $\chi(\mathbf{L}, f, t)$ and of its unit vector (6.5):

$$\mathbf{U}^0 \approx \mathbf{R}^0 + \frac{\sigma_2 - \sigma_1}{2\sigma|\mathbf{R}|}[\mathbf{d} - (\mathbf{d}^T\mathbf{R}^0)\mathbf{R}^0] \tag{6.21}$$

where σ is the total target RCS

$$\sigma = \sigma_1 + \sigma_2 + 2\sqrt{\sigma_1\sigma_2}\cos\left(\frac{4\pi}{\lambda}\mathbf{d}^T\mathbf{R}^0\right) \tag{6.22}$$

From (6.21) one can evaluate the angular glint vector

$$\mathbf{U}^0 - \mathbf{R}^0 = \frac{\sigma_2 - \sigma_1}{2\sigma|\mathbf{R}|}[\mathbf{d} - (\mathbf{d}^T\mathbf{R}^0)\mathbf{R}^0] = \frac{\sigma_2 - \sigma_1}{2\sigma|\mathbf{R}|}\mathbf{d}_{\text{trans}} \tag{6.23}$$

where $\mathbf{d}_{\text{trans}}$ is the transversal component of vector $\mathbf{d}$, which is normal to vector $\mathbf{R}^0$.

The calculation method presented above for angular glint will be used for other glints except that the partial derivatives with respect to frequency f or time t will replace the partial derivative with respect to vector $\mathbf{L}$.

Range Glint Evaluation. The apparent range entered into (6.11) was defined through the partial frequency derivative of the phase distribution function $\chi(0, f, t) = \mathrm{atan}[B(0)/A(0)]$ of the arbitrary nonsinusoidal signal. The partial frequency derivative $\partial\chi/\partial f$ is determined by an equation analogous to (6.18), but the partial derivatives with respect to variable $\mathbf{L}$ are replaced here by the partial derivatives with respect to frequency f. Using the values (6.15) for $\mathbf{L} \to 0$ and approximation (6.16), we obtain

$$\frac{\partial B(0)}{\partial f} \approx \frac{4\pi}{c}|\mathbf{R}|\left[A(0) - \frac{\mathbf{d}^{\mathrm{T}}\mathbf{R}^0}{2|\mathbf{R}|}A_1(0)\right], \qquad (6.24)$$

$$\frac{\partial A(0)}{\partial f} \approx -\frac{4\pi}{c}|\mathbf{R}|\left[B(0) - \frac{\mathbf{d}^{\mathrm{T}}\mathbf{R}^0}{2|\mathbf{R}|}B_1(0)\right]$$

According to (6.11), the range glint is defined by the scalar value $\mathbf{d}^{\mathrm{T}}\mathbf{R}^0$ of the radial component of vector $\mathbf{d}$:

$$\Delta R(t) \approx \frac{\sigma_2 - \sigma_1}{2\sigma}\mathbf{d}^{\mathrm{T}}\mathbf{R}^0 \qquad (6.25)$$

Spatial Glint Evaluation and Discussion. In accordance with (6.12), (6.23), and (6.25), the spatial glint is determined by

$$\mathbf{a} \approx \frac{\sigma_2 - \sigma_1}{2\sigma}\mathbf{d} \qquad (6.26)$$

Target apparent position is situated according to (6.26) on the direct line connecting the target elements. Glint is absent ($\mathbf{a} = 0$) if $\sigma_2 = \sigma_1$ and $\sigma \neq 0$. If $\sigma_2 \gg \sigma_1$ or $\sigma_2 \ll \sigma_1$, the apparent target position coincides with the most intense of the target elements. If the values σ_2 and σ_1 are near to each other, the RCS σ can achieve its minimum value $(\sqrt{\sigma_2} - \sqrt{\sigma_1})^2$. The glint achieves then its maximum value $\mathbf{a} = (\sqrt{\sigma_2} + \sqrt{\sigma_1})\mathbf{d}/(\sqrt{\sigma_2} - \sqrt{\sigma_1})$, which can be much greater than the distance of a target's elements from its center. Both the spatial glint $\mathbf{a}$ and RCS σ depend on the angle θ between the vectors $\mathbf{d}$ and $\mathbf{R}^0$. Examples of dependencies of the spatial glint on angle θ (solid lines) and analogous dependencies of RCS (dotted line) are shown in Figure 6.6 for the ratios of σ_1/σ_2 equal to 1.1 [Figure 6.6(a)] and 2.0 [Figure 6.6(b)] and the ratio $|\mathbf{d}|/\lambda = 1$. Both the RCS σ and space glint $|\mathbf{a}|$ are presented in logarithmic scale. For the convenience of demonstration, the dependencies of Figure 6.6 correspond to the meter waveband.

Doppler Frequency Glint Evaluation and Discussion. Entered into (6.13), apparent doppler frequency was defined through the partial time derivative of the phase distribution function $\chi(0, f, t) = \mathrm{atan}[B(0)/A(0)]$. Under the condition of uniform translational target motion, let us substitute the value $\mathbf{R}(t) = \mathbf{R} - \mathbf{v}t$ for $\mathbf{R}$. The partial derivative $\partial\chi/\partial t$ is defined again by the equation analogous to (6.18), but partial derivatives with respect to variable $\mathbf{L}$ are replaced by the partial derivatives with respect to time t, expressed through the derivatives of delays with respect to t:

$$\frac{\partial\Delta t_{1,2}(0)}{\partial t} = \frac{\partial\Delta t_{1,2}(0)}{\partial\mathbf{R}}\frac{d\mathbf{R}}{dt} = \frac{2(\mathbf{R} \mp \mathbf{d}/2)^{\mathrm{T}}\mathbf{v}}{c|\mathbf{R} \mp \mathbf{d}/2|} \tag{6.27}$$

$$\approx \frac{2}{c}\left[\mathbf{R}^0\left(1 \pm \frac{\mathbf{d}^{\mathrm{T}}\mathbf{R}^0}{2|\mathbf{R}|}\right) \mp \frac{\mathbf{d}}{2|\mathbf{R}|}\right]^{\mathrm{T}}\mathbf{v}$$

Computing as before the partial derivatives $\partial A(0)/\partial t$ and $\partial B(0)/\partial t$, substituting them into the equation analogous to (6.18), and using (6.13), one obtains

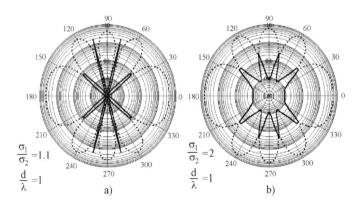

Figure 6.6 Dependencies of value of spatial glint $|\mathbf{a}|$ (solid lines) and of RCS (dotted lines) in logarithmic scale for the two-element target for the ratios of σ_1/σ_2 equal to (a) 1.1 and (b) 2.0, and the ratio $|\mathbf{d}|/\lambda = 1$.

$$\Delta F_D \approx \frac{\sigma_2 - \sigma_1}{\sigma |\mathbf{R}|} \frac{\mathbf{d}_{trans}^T \mathbf{v}}{\lambda} = \frac{2}{\lambda}(\mathbf{U}^0 - \mathbf{R}^0)^T \mathbf{v}$$

This equation shows that the doppler glint is connected with angular glint (6.23) and is caused by the transversal component of the target velocity.

6.2.3 Possible Simplification of Angular Glint Simulation for Real Targets and Optimal Radar

Equation (6.23) defines the angular glint only for the two-element target model. To simulate the angular glint for the complex targets, one can use the relation (6.5), where the increment $d\mathbf{L} \to 0$ of only the receiving position is included.

If such increment is not provided by the simulation program, one can consider the small increments $d\mathbf{R}$ of the transmitting-receiving position. Let us consider the signal delay corresponding to ith bright point $\Delta t_i = [|\mathbf{R}_i| + |\mathbf{R}_i - \mathbf{L}|]/c$, where $\mathbf{R}_i = \mathbf{R} + \boldsymbol{\rho}_i$ is its radius-vector in the radar coordinate system, $\mathbf{R}$ is the radius-vector of the origin of the target coordinate system in the radar system, and $\boldsymbol{\rho}_i$ is the radius-vector of ith bright point in the target coordinate system. Let us compare the derivatives of the delay Δt_i as the functions of $\mathbf{L}$ and of $\mathbf{R}$ for $\mathbf{L} \to 0$:

$$\frac{d\Delta t_i}{d\mathbf{L}}\bigg|_{\mathbf{L} \to 0} = -\frac{|\mathbf{R}_i|}{c} = -\frac{1}{2}\frac{d\Delta t_i}{d\mathbf{R}}\bigg|_{\mathbf{L} \to 0}$$

Using the result of this comparison, let us change the derivative of phase front function $\chi(\mathbf{L}, f, t) = \arg \dot{E}(\mathbf{L}, f, t)$ entered into (6.5) for $\mathbf{L} \to 0$:

$$\frac{d\chi}{d\mathbf{L}} = \sum_i \frac{\partial \chi}{\partial \Delta t_i}\frac{d\Delta t_i}{d\mathbf{L}} = -\frac{1}{2}\sum_i \frac{\partial \chi}{\partial \Delta t_i}\frac{d\Delta t_i}{d\mathbf{R}} = -\frac{1}{2}\frac{d\chi}{d\mathbf{R}} \tag{6.28}$$

We see that a displacement of receiving position by a small vector $d\mathbf{L}$ can be replaced by the halved displacement of transmitting-receiving position $d\mathbf{R} = -d\mathbf{L}/2$.

As it was explained with Figure 6.3(b) above, the glint $\Delta\beta$ is defined by the vector $\mathbf{U}^0 - \mathbf{R}^0$. This vector is collinear to the unit vector $\mathbf{R}_{transv}^0$ transversal to $\mathbf{R}^0$ and is equal to $\Delta\beta\mathbf{R}_{transv}^0$, so that $\Delta\beta\mathbf{R}_{transv}^0 = \mathbf{U}^0 - \mathbf{R}^0$, $\Delta\beta(\mathbf{R}_{transv}^0)^T\mathbf{R}_{transv}^0 = (\mathbf{R}_{transv}^0)^T(\mathbf{U}^0 - \mathbf{R}^0)$ or $\Delta\beta = (\mathbf{R}_{transv}^0)^T\mathbf{U}^0$.

Using (6.5) and (6.28), we obtain

$$\Delta\beta = (\mathbf{R}_{\text{transv}}^{0})^{\text{T}} \frac{d\chi}{d\mathbf{R}} \bigg/ \left| \frac{d\chi}{d\mathbf{R}} \right| \tag{6.29}$$

where the vector derivative of χ (gradient χ) is

$$\frac{d\chi}{d\mathbf{R}} = \frac{\partial\chi}{\partial R}\mathbf{R}^{0} + \frac{\partial\chi}{\partial R_{\text{transv}}}\mathbf{R}_{\text{transv}}^{0} = \frac{\partial\chi}{\partial R}\mathbf{R}^{0} + \frac{1}{R}\frac{\partial\chi}{\partial\beta}\mathbf{R}_{\text{transv}}^{0} \tag{6.30}$$

The simulation program allows estimating the reflected signal phase $\chi = \chi(\beta)$ via target azimuth β, so that partial derivative $\partial\chi/\partial\beta$ can be found from simulation. The partial derivative $\partial\chi/\partial R = \partial[4\pi R/\lambda + \text{const}]/\partial R = 4\pi/\lambda \gg R^{-1}\partial\chi/\partial\beta$. From (6.29) and (6.30), we obtain

$$R\Delta\beta = \frac{\lambda}{4\pi} \frac{\partial\chi/\partial\beta}{\sqrt{1 + (\lambda/4\pi R)^{2}(\partial\chi/\partial\beta)^{2}}} \approx \frac{\lambda}{4\pi}\partial\chi/\partial\beta \tag{6.31}$$

The last approximation is correct outside the Fresnel zone. In this case the azimuth derivative $\partial\chi/\partial\beta$ can be replaced by the aspect derivative $\partial\chi/\partial\psi$, which can simply be simulated.

6.2.4 Simulation Examples for Real Targets and Radar

Target angular glint can be considered both for sequential scan type and monopulse type angle measurements, as well as for single and integrated measurements in view of target motion and noise. But simulation examples are given below for a single measurement without thermal noise, 10-cm wavelength, 6m receiving aperture width, 10-km (about 5.4 miles) target range, 0° to 20° aspect angle sector, and high pulse repetition frequency. The reception using a four-element antenna was simulated.

Examples of simulated envelopes of pulse trains for angle measurement of the sequential scan type with large radar-target contact time are shown in Figure 6.7. Figure 6.7(a) corresponds to the ALCM missile, and Figure 6.7(b) corresponds to the AH-64 helicopter. It is seen how the fluctuations of the pulse train interfere with the sequential scan type measurement.

Examples of simulated angular error pdf are shown in Figure 6.8 for Tu-16 and An-26 aircraft, AH-64 helicopter, and ALCM missile. Figure 6.8(a) corresponds to sequential scan type measurement, and Figure 6.8(b)

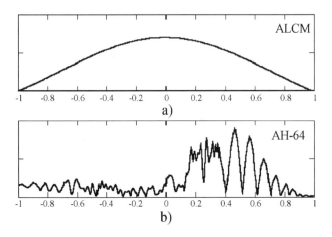

Figure 6.7 Simulated envelopes of the pulse trains for (a) ALCM missile and (b) AH-64 helicopter determining the errors in the angle measurement of the sequential scan type.

corresponds to monopulse. The scale of transverse range introduced in the last case generalizes the simulation's results.

Examples of simulated pdf of range error are shown in Figure 6.9 for enumerated targets and their illumination by the simple rectangular pulse of 1 μs duration. The result can be generalized on various simple pulse form and duration.

The shape of given curves is influenced by the following factors: method of coordinate measurement, target effective dimensions and engine placement, parameters of turbine or propeller modulation, and pulse repetition frequency.

6.3 Some Aspects of the Wideband Signal Use in Detection and Tracking

Wideband signals can be used not only in recognition, but also in detection and tracking. It has long been known that such signals can improve capabilities of detection and tracking under conditions of clutter and jamming [21]. However, there is sometimes reluctance to exploit these advantages due to the complication of signal processing, aggravation of electromagnetic compatibility, decrease in radar range, and accuracy of tracking of multi-element targets in some scattered instances.

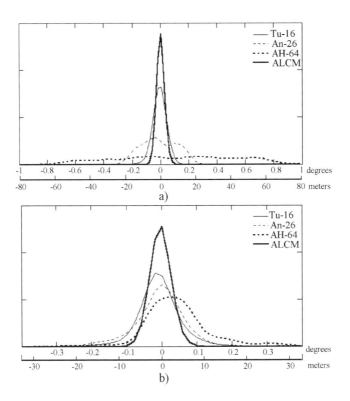

Figure 6.8 Simulated pdf of angular and cross-range errors for (a) sequential scan–type and (b) monotype measurements for various targets at a range of 5 km.

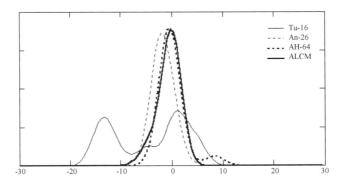

Figure 6.9 Simulated pdf of range measurement error for various targets with an NB signal of 1-MHz bandwidth.

Nowadays, the progress in digital processing allows us to consider seriously the use of wideband detection and tracking of targets. The electromagnetic compatibility problems must be thoroughly considered in each special case. It is not evident that narrowband signals with the great spectral density of illumination should always be preferred to wideband signals with small spectral density. Time selection can supplement frequency not only in narrowband systems, but in wideband systems too.

According to the results of simulation, a definite increase in signal bandwidth can increase (not decrease) the target detection probabilities (Section 6.3.1) and decrease (not increase) the glint effects in measurement (Section 6.3.2) and tracking (Section 6.3.3).

6.3.1   Simulation of Target Detection with Wideband Signals

Statement of the Simulation Problem. Three variants of detection were considered: narrowband detection, wideband cumulative detection, and wideband detection with noncoherent integration using the "log-scale" scheme. The variant of wideband cumulative detection corresponds to the absence of any scheme of noncoherent integration within the RP. The target is detected if at least one of N elements of its RP is above the detection threshold. Probability of detection D for this case can be expressed with unequal probability D_n of detection of various target elements as

$$D = 1 - \prod_{n=1}^{N} (1 - D_n).$$

The introduction of special schemes for noncoherent integration is, in principle, a complex enough task. It would be desirable to take into account the variety of target types, their aspect sectors, and their specific orientation within the sector. Since accounting for all these factors is hardly possible, a simplified detection scheme was chosen, which we called the "log-scale" one. The RP variants were approximated by rectangles, and the range extent of these rectangles was as varied as the product of the range resolution and the powers of a definite number, particularly by the powers 2^m, $m = 0, \ldots,$ M. After noncoherent signal integration in $M + 1$ channels, obtained in such a manner, the outputs of these channels were compared with unequal thresholds, which were set in order to make false alarm rates equal for all the channels. Having been compared with the thresholds, the outputs were subjected to logical processing by the criterion "one from $M + 1$" (Figure 6.10). The "log-scale" scheme with six channels was chosen for simulation, and the false alarm rate in each channel was set to $F_0 = 10^{-4}$, so that the

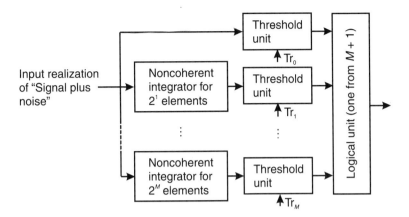

Figure 6.10 The "log-scale" scheme for the WB signal detection.

simulated false alarm rate at the output of the logical unit constituted $F = 5.7 \cdot 10^{-4} \approx 6 \times 10^{-4}$.

Simulation Results and Discussion. Figure 6.11 shows the simulated detectability factors for the Tu-16 aircraft corresponding to the variants of detection listed above for the equal false alarm rate of illustrative level $F = 5.7 \times 10^{-4}$ per resolution element. The wavelength was $\lambda = 5$ cm. In the whole the range gate of 96 elements was processed. Three thousand realizations of "signal plus noise" were used for evaluation of each point of the curves. These realizations corresponded to target nose-on course aspects from $0°$ to

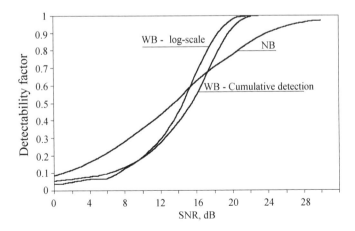

Figure 6.11 Detectability factors simulated for Tu-16 aircraft using NB and WB illumination for the "log-scale" and cumulative detection schemes.

20°, to target roll aspects from −5° to 5°, and to target pitch aspects from −1° to 9°. Wideband and narrowband detection was simulated using chirp deviations of 80 MHz and 1 MHz, respectively. There is a gain in signal-to-noise ratio when the target is detected with a high detectability factor using a wideband signal instead of a narrowband signal. It can be seen from Figure 6.11 that if detection probability is $D = 0.85$, this gain is about 4 dB for cumulative detection and 5 dB for the "log-scale" detection scheme.

6.3.2   Simulation of Target Range Glint in a Single Wideband Measurement

Range resolution of the target elements reduces the range glint, which will be determined by the RPs. If the signal is intense enough, then angular glint will be reduced too, if the angular measurements are performed for several range cells and their results are averaged. The results for the range glint simulation are given below.

Statement of the Range Glint Simulation Problem. Range measurement for the Tu-16 aircraft was simulated using a narrowband chirp signal of 1 MHz and a wideband chirp signal of 80 MHz deviation. In the latter case, the range was measured by (1) maximum sample of the RP and (2) median sample of the RP. Measured values of range were then compared to actual values set in the model in order to compute the range measurement error. The test measurements were carried out 500 times for various conditions. As a result, the pdf of range measurement error was estimated.

Results of the Range Glint Simulation and Discussion. Figure 6.12 shows the simulated pdf of range shift of the target effective center of scattering (range glint) for the 1-MHz narrowband signal (curve 1), and for the 80-MHz wideband signal with range measurement by the maximum RP sample (curve 2), and by the median RP sample (curve 3). It can be seen that the largest dispersion of range estimates between their edge values is for the narrowband measurement. For the wideband signal, it decreases even if the range is measured using the RP sample with maximum amplitude. This dispersion is of the lowest magnitude if the wideband range measurement is performed using the median sample of RP.

    The glint errors can be compared with the thermal noise errors. The potential root-mean-square range error due to noise was evaluated as $\sigma_{range} = c\sigma_{time}/2$, where c is the light velocity in free space. The value of time error σ_{time} was calculated by the Woodward formulae $\sigma_{time} = 1/\beta\sqrt{2E/N_0}$ [20].

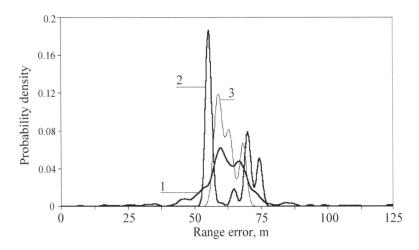

Figure 6.12 The pdf of range glint simulated for Tu-16 aircraft illuminated by the 1-MHz NB (curve 1), the 80-MHz WB signal for evaluation of the maximum (curve 2), and the median (curve 3) sample of the RP.

Here, β is the bandwidth, and E/N_0 is the energy signal-to-noise ratio. For the ratio equal to 15 dB or $E/N_0 \approx 31$, the potential error due to noise is about 19m for the signal bandwidth of 1 MHz and about 0.24m for the signal bandwidth of 80 MHz.

6.3.3 Simulation of Target Range Glint in Wideband Tracking

The choice of the tracking system is not very significant in our case. We used the simple discrete tracking system without the exponential smoothing synthesized on the basis of the model of movement of the material point with ideally constant radial velocity. After obtaining the current range estimates $\hat{r}_k$ ($k = 0, 1, \ldots, N - 1$) with errors of the stationary noise type, the tracking system computes more precise estimates of range $\hat{R}_{k+1}$ for their numbers $k = 0, \ldots, N - 1$ and of radial velocity $\hat{V}_{k+1}$ for the numbers $k = 1, \ldots, N - 1$

$$\left\| \begin{matrix} \hat{R}_{k+1} \\ \hat{V}_{k+1} \end{matrix} \right\| = \left\| \begin{matrix} \hat{R}_k + \hat{V}_k T \\ \hat{V}_k \end{matrix} \right\| + \left\| \begin{matrix} 2(2k + 1)/(k + 1)(k + 2) \\ 6/(k + 1)(k + 2)T \end{matrix} \right\| (\hat{r}_{k+1} - \hat{R}_k - \hat{V}_k T)$$

(6.32)

From here, one can find: (1) for $k = 0$, that $\hat{R}_1 = \hat{r}_1$; (2) for $k = 1$, that $\hat{R}_2 = \hat{r}_2$ and $\hat{V}_2 = (\hat{r}_2 - \hat{r}_1)/T$; (3) for $k = 2$, that $\hat{R}_3 = \hat{R}_2 + \hat{V}_2 T + \frac{5}{6}(\hat{r}_3 - \hat{R}_2 - \hat{V}_2 T)$ and $\hat{V}_3 = \hat{V}_2 + \frac{1}{2T}(\hat{r}_3 - \hat{R}_2 - \hat{V}_2 T)$, etc.

Statement of the Simulation Problem.   The tracking system was provided with N = 20 simulated current range estimates $\hat{r}_k$ of the Tu-16 aircraft. The aircraft was piloted along the line-of-sight with a velocity of 150 m/s under conditions of clear weather turbulence. The atmosphere turbulence led to angle yaws of the aircraft with standard deviation of about 1.5°. At the output of the tracking system, the smoothed estimates of range $\hat{R}_k$ and velocity $\hat{V}_k$ of target were obtained. The errors in range and velocity measurements for each flight path were calculated comparing these smoothed estimates with the actual range and velocity. The errors were simulated for a radar operated at λ = 5 cm with signals of 1- and 80-MHz bandwidth and averaged for 40 various flight paths. Simulation was carried out without introducing the receiver thermal noise.

Simulation Results and Their Discussion.   Figure 6.13 shows the error standard deviations in narrowband and wideband range measurements for the tracking algorithm (6.29) calculated for the case of noise absence. It can be seen that the range measurement error for the wideband signal is much lower than for the narrowband signal.

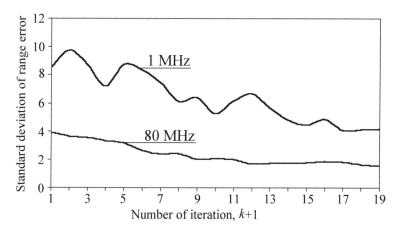

Figure 6.13   Standard deviations of the range errors at the output of tracking system simulated for Tu-16 aircraft using NB (1 MHz) and WB (80 MHz) target illumination.

References

[1]   Marcum, J. I., "Statistical Theory of Target Detection by Pulsed Radar," *IEEE Trans.*, IT-6, April 1960, pp. 145–268.

[2]   Swerling, P., "Probability of Detection for Fluctuating Targets," *IEEE Trans.*, IT-6, April 1960, pp. 269–308.

[3]   Swerling, P., "Detection of Fluctuating Pulsed Signals in the Presence of Noise," *IRE Trans.*, IT-3, No. 3, September 1957, pp. 175–178.

[4]   Weinstock, W. W., "Radar Cross Section Target Models." In *Modern Radar*, Chapter 5, R. S. Berkowitz, (ed.), New York: Wiley, 1965.

[5]   Meyer, D. P., and H. A. Mayer, *Radar Target Detection—Handbook of Theory and Practice*, New York: Academic Press, 1973.

[6]   Heidebreder, G., and R. Mitchell, "Detection Probabilities for Log-Normally Distributed Signals," *IEEE Trans.*, AES-3, No. 3, 1967, pp. 5–13.

[7]   Shnidman, D. A., "Radar Detection Probabilities and their Calculation," *IEEE Trans.*, AES-31, No. 3, July 1995, pp. 928–950.

[8]   Johnston, S. L., "Target Fluctuation Models for Radar System Design and Performance Analysis: An Overview of Three Papers," *IEEE Trans.*, AES-33, No 2, April 1997, pp. 696–697.

[9]   Swerling, P., "Radar Probability of Detection for Some Additional Fluctuating Target Cases," *IEEE Trans.*, AES-33, No. 2, April 1997, pp. 698–709.

[10] Xu, X., and P. Huang, "A New RCS Statistical Model of Radar Targets," *IEEE Trans.*, AES-33, No. 2, April 1997, pp. 710–714.

[11] Johnston, S. L., "Target Model Pitfalls (Illness, Diagnosis, and Prescription)," *IEEE Trans.*, AES-33, No. 2, April 1997, pp. 715–720.

[12] Shirman, Y. D. et al., "Study of Aerial Target Radar Recognition by Method of Backscattering Computer Simulation," *Proc. Antenna Applications Symp.*, Allerton Park Monticello, IL, September 1999, pp. 431–447.

[13] Shirman, Y. D. et al., "Aerial Target Backscattering Simulation and Study of Radar Recognition, Detection and Tracking," *IEEE Int. Radar-2000*, Washington, DC, May 2000, pp. 521–526.

[14] Barton, D. K., *Modern Radar System Analysis*, Norwood, MA: Artech House, 1988.

[15] Delano, R. H., "A Theory of Target Glint or Angular Scintillation in Radar Tracking," *Proc. IRE 41*, No. 8, December 1953, pp. 1778–1784.

[16] Shirman, Y. D., and V.N. Golikov, "To the Theory of the Scattering Effective Center Walk," *Radiotekhnika i Electronika*, Vol. 13, November 1968, pp. 2077–2079 (in Russian).

[17] Shirman, Y. D. (ed.), *Theoretical Foundations of Radar*, Moscow: Sovetskoe Radio Publishing House, 1970; Berlin: Militärverlag, 1977

[18] Razskazovsky, V. B., "Statistical Characteristics of Group Delay of Scatterers' Set," *Radiotekhnika i Electronika*, Vol. 16, November 1971, pp. 2105–2109 (in Russian).

[19]   Ostrovityanov, R. V., and F. A. Basalov, "Statistical Theory of Extended Radar Targets," Moscow: Sovetskoe Radio Publishing House, 1982; Norwood, MA: Artech House, 1985.

[20]   Woodward, P. M., *Probability and Information Theory with Applications to Radar*, Oxford: Pergamon, 1953; Norwood, MA: Artech House, 1980.

[21]   Shirman, Y. D., *Resolution and Compression of Signals*, Moscow: Sovetskoe Radio Publishing House, 1974 (in Russian).

7

Some Expansions of the Scattering Simulation

Expanding on the issues considered in previous chapters, we discuss here the scattering of radio waves by targets (1) with imperfectly conducting surfaces having absorbing coating, especially; and (2) in bistatic radar as outlined in Section 1.3.

We consider in this chapter an approximate scattering analysis at high frequencies using an augmented variant of physical optics. The augmentation consists of including both perfectly and imperfectly conducting surfaces. The boundary between the light and shadow zones is called the terminator. As usual, in physical optics approximations [1–6], the terminator is supposed to be sharp, and unlike [7] the half-shadow (penumbra) zone is not accounted for. But the shadow radiation [5, 6] is included in the concept of physical optics as necessary for approximate consideration of bistatic systems.

The geometric model of bistatic scattering is shown in Figure 7.1(a). The distance from the radar transmitter to the origin of the target coordinate system is denoted by R, and the distance from this origin to the radar receiver is denoted by r. The angle α between the directions from target to transmitter and receiver is known as the bistatic angle. Only acute bistatic angles and angles near to 180° will be considered. The lines (surfaces) of constant range sum $R + r$ are shown in Figure 7.1(b). They are ellipses (ellipsoids) with transmitter and receiver as their focuses. The normal to each ellipse (ellipsoid) is the bisector of the bistatic angle. The sum of unit vectors $\mathbf{R}^0 + (-\mathbf{r}^0) = \mathbf{R}^0 - \mathbf{r}^0$ is directed along this bisector, being the gradient of the range sum.

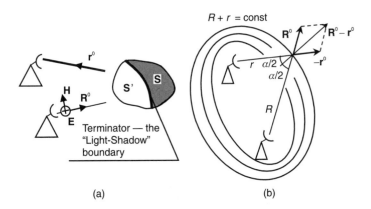

Figure 7.1 Geometric model of (a) bistatic scattering and (b) ellipses of constant range sum.

For large R and r the part of the ellipsoid surface near the target can be regarded as a plane of constant range sum.

In this chapter the physical optics approximation is carried out without separating the target's simplest (specular, bright) components. The latter procedure simplified all the calculations of previous chapters and eased the accounting for the target's motion and the signal's rotational modulation. As with other high frequency approximations, this separation is not the most precise of possible approximations. The separation would be exact if secondary radiation of all Fresnel zones neighboring the bright element were compensated completely. Incomplete compensation observed at low frequencies leads to errors. We do not claim that the methods used in Chapter 7 will always provide better results if the frequency is decreased significantly. On the contrary, the method of bright elements can simplify the accounting for rereflections, although they can be accounted for also by stricter methods [8]. The method of specular elements is also used below. We suggest in Section 7.1.7 that the majority of methods considered in this chapter is best applied in a definite frequency domain.

In Section 7.1 the scattering effects for stationary (monochromatic) illumination of targets will be considered. In Section 7.2 we consider the nonstationary case and introduce and clarify the high frequency impulse response (HFIR) and unit step response (HFUSR) of targets.

7.1 Scattering Effects for Stationary (Monochromatic) Illumination of Targets

The aim of this section is to describe some effective methods of calculating RCS for the targets and their parts with perfectly and imperfectly conducting surfaces observed by monostatic and bistatic radar. In Sections 7.1.1 through 7.1.6 we consider targets both uncovered and covered with absorber, and which are large in comparison with wavelength. Approximate asymptotic methods of physical optics are used here. In Section 7.1.7 we consider the parts of targets with discontinuities covered by absorbers, and methods of physical optics are combined with the methods of eigenfunctions.

7.1.1 Expressions of Scattered Field for Targets with Perfectly Conducting Surfaces

The problem of scattering will be divided artificially into inducing currents on the perfectly conducting surface of a target and radiation of the surface currents, which can be considered then as extraneous. The target is supposed to be:

- Approximated by a smooth closed convex surface $S(\boldsymbol{\rho})$, where $\boldsymbol{\rho} = \|\rho_1, \rho_2, \rho_3\|^{\mathrm{T}}$ is the radius vector of a point of the surface;
- Illuminated by the plane sinusoidal wave with intensities of magnetic $\mathbf{H}_{\mathrm{tg}}$ and electric $\mathbf{E}_{\mathrm{tg}}$ fields in its surrounding, so that

$$\mathbf{H}_{\mathrm{tg}}(\boldsymbol{\rho}) = \mathbf{H}_0 \exp(-jk((\mathbf{R}^0)^{\mathrm{T}}\boldsymbol{\rho} + R)), \quad \mathbf{E}_{\mathrm{tg}}(\boldsymbol{\rho}) = \sqrt{\frac{\mu_0}{\epsilon_0}}\mathbf{H}_{\mathrm{tg}}(\boldsymbol{\rho}) \times \mathbf{R}^0$$

$$(7.1)$$

Here, $\mathbf{H}_0 = \mathbf{H}_{\mathrm{tg}}(0)$ is the value of $\mathbf{H}_{\mathrm{tg}}(\boldsymbol{\rho})$ for $\boldsymbol{\rho} = 0$, $k = 2\pi f/c$ is the wave number in free space, $\mathbf{R}^0$ is the unit vector in the propagation direction, R is the distance between the illumination source and origin O_{tg} of the coordinate system associated with the target, μ_0 and ϵ_0 are the permeability and permittivity of free space, and $\sqrt{\frac{\mu_0}{\epsilon_0}} = Z_0$ is its wave impedance. In the classical approximation of physical optics, the surface currents are exited on the illuminated side of a target only (Figure 7.1). In this approximation the target can be divided into elements with dimensions that are large compared to wavelength λ. The illumination causes wave reflection and surface current with the density $\mathbf{K} = 2\mathbf{n} \times \mathbf{H}_{\mathrm{tg}} = \mathbf{n} \times 2\mathbf{H}_{\mathrm{tg}}$, where $\mathbf{n} = \mathbf{n}(\boldsymbol{\rho})$ is the unit vector of an internal normal to S. This solves the first part of our problem.

In the second part of the problem we abstract ourselves from the first part. The doubled field $\mathbf{H}_{\mathrm{extr}} = 2\mathbf{H}_{\mathrm{tg}}$ and corresponding surface density of current $\mathbf{K} \approx 2\mathbf{n} \times \mathbf{H}_{\mathrm{tg}}$ we consider as extraneous sources of radiation $\mathbf{H}_{\mathrm{extr}} \approx 2\mathbf{H}_{\mathrm{tg}}$ and $\mathbf{K}_{\mathrm{extr}} \approx \mathbf{n} \times \mathbf{H}_{\mathrm{extr}}$. Starting from the known equation of radiation from an elementary dipole in the far zone, we obtain the field at the remote reception point at distance r from the target in the propagation direction $\mathbf{r}^0$:

$$\mathbf{H}_{\text{rec}} \approx jk\frac{\exp(-jkr)}{4\pi r}\int\limits_{S}[\mathbf{K}_{\text{extr}}\times\mathbf{r}^0]\exp(-jk((\mathbf{r}^0)^\mathsf{T}\boldsymbol{\rho}))dS \qquad (7.2)$$

$$\mathbf{E}_{\text{rec}} = \sqrt{\frac{\mu_0}{\epsilon_0}}\,\mathbf{H}_{\text{rec}}\times\mathbf{r}^0$$

where $\mathbf{K}_{\text{extr}} \approx 2(\mathbf{n}\times\mathbf{H}_{\text{tg}})$ for the "illuminated" part S_1 of target surface. It is assumed for small bistatic angles that $\mathbf{K}_{\text{extr}} \approx 0$ for the nonilluminated part of the target surface. But it will be shown in Section 7.1.7 that for large bistatic angles, especially those approaching $180°$, one has to account in calculations for the so-called "shadow radiation."

7.1.2 Expressions of Scattered Field for Targets with Imperfectly Conducting Surfaces

Duality of the Solutions of Maxwell's Equations. Let us consider Maxwell's equations for the media without sources of electromagnetic field:

$$\nabla\times\mathbf{E} = -\partial(\mu_r\mu_0\mathbf{H})/\partial t, \quad \nabla\times\mathbf{H} = \partial(\epsilon_r\epsilon_0\mathbf{E})/\partial t \qquad (7.3)$$

where μ_r is the relative permeability of media, μ_0 is the absolute permeability of free space, ϵ_r is relative permittivity of media, and ϵ_0 is absolute permittivity of free space. If we replace the $\mathbf{E}$, $\mathbf{H}$, μ_r, μ_0, ϵ_r, ϵ_0 by $-\mathbf{H}$, $\mathbf{E}$, ϵ_r, ϵ_0, μ_r, μ_0 correspondingly, the first and the second of these equations will become the second and first equations, respectively [3].

It means that such a replacement in one of the solutions (7.2) of Maxwell's equations leads to another solution

$$\mathbf{E}_{\text{rec}} \approx jk\frac{\exp(-jkr)}{4\pi r}\int\limits_{S}[(\mathbf{n}\times\mathbf{E}_{\text{extr}})\times\mathbf{r}^0]\exp(-jk((\mathbf{r}^0)^\mathsf{T}\boldsymbol{\rho})dS, \qquad (7.4)$$

$$\mathbf{H}_{\text{rec}} = -\sqrt{\frac{\epsilon_0}{\mu_0}}\,\mathbf{E}_{\text{rec}}\times\mathbf{r}^0$$

where $\sqrt{\dfrac{\epsilon_0}{\mu_0}} = Y_0$ is the wave admittance of free space.

Scattered Fields for Targets with Imperfectly Conducting Surfaces. Owing to the linearity of Maxwell equations (7.4), the superposition of solutions also satisfies them. Therefore, one can use also the superposition of solutions (7.2) and (7.4):

$$\mathbf{H}_{rec} \approx jk \frac{\exp(-jkr)}{4\pi r} \int_S [\mathbf{J}(\mathbf{n}, \mathbf{r}^0) \times \mathbf{r}^0] \exp(-jk((\mathbf{r}^0)^T \boldsymbol{\rho}) dS, \qquad (7.5)$$

$$\mathbf{E}_{rec} = \sqrt{\frac{\mu_0}{\epsilon_0}} \mathbf{H}_{rec} \times \mathbf{r}^0$$

where $\mathbf{J}(\mathbf{n}, \mathbf{r}^0)$ is known as the Huygens elementary radiator [3, 4], which replaces the density of extraneous currents $\mathbf{K}_{extr} = \mathbf{n} \times \mathbf{H}_{extr}$ in (7.2)

$$\mathbf{J}(\mathbf{n}, \mathbf{r}^0) = \mathbf{n} \times \mathbf{H}_{extr} - \sqrt{\frac{\epsilon_0}{\mu_0}} [\mathbf{n} \times \mathbf{E}_{extr}] \times \mathbf{r}^0 \qquad (7.6)$$

Equations (7.5) and (7.6) will be applied below to the solutions of scattering problems, mostly in the classical approximation of physical optics. Using the result (7.5) in this approximation we suppose that $\mathbf{J}(\mathbf{n}, \mathbf{r}^0) \neq 0$ for the "illuminated" part S_1 of the target surface and $\mathbf{J}(\mathbf{n}, \mathbf{r}^0) \approx 0$ for its "nonilluminated" part. To obtain examples of sources $\mathbf{n} \times \mathbf{H}_{extr}$ and $\mathbf{n} \times \mathbf{E}_{extr}$ causing radiation, let us consider the propagation of plane waves in parallel layers [1] of uniform and isotropic media including absorbing layers.

7.1.3 The Plane Waves in Parallel Uniform Isotropic Infinite Layers

The Case of Two Layers. This case corresponds to the straight-line propagation of the incident wave in dielectric media, its specular reflection from the interface between the layers, and refraction through interface. Figure 7.2(a) corresponds to an **E** field polarized perpendicularly to the plane of incidence; Figure 7.2(b) corresponds to an **E** field polarized in parallel to this plane. In both cases the angles of reflection are equal to those of incidence $\theta_\perp = \theta_\parallel = \theta_1$, and the angles of refraction θ_2 and of incidence θ_1 are related according to the Snell's law

$$\sin\theta_2 = \dot{\nu}\sin\theta_1, \quad \dot{\nu} = \sqrt{\dot{\mu}_1 \dot{\epsilon}_1 / \dot{\mu}_2 \dot{\epsilon}_2} \qquad (7.7)$$

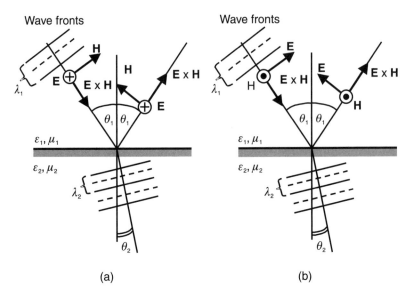

Figure 7.2 Reflection and refraction at an interface of two infinite dielectric layers for two polarizations of an **E** field: (a) perpendicular to the plane of incidence, and (b) parallel to it.

where $\dot{\mu}_1$, $\dot{\mu}_2$, $\dot{\epsilon}_1$, $\dot{\epsilon}_2$ are the complex relative permeabilities and permittivities of layers; their imaginary parts account for energy absorption.

The boundary conditions are the equality of the tangential components of the resulting **E** and **H** fields at the interface. Under the assumption of unit field intensity $|\mathbf{E}_1| = 1$ of the incident wave, the boundary conditions for these cases (Figure 7.2) have the form

$$(a) \; 1 + \dot{R}_{\perp} = \dot{R}_{\perp\text{refr}}$$

$$\sqrt{\frac{\dot{\epsilon}_1 \epsilon_0}{\dot{\mu}_1 \mu_0}}(1 - \dot{R}_{\perp}) \cos\theta_1 = \sqrt{\frac{\dot{\epsilon}_2 \epsilon_0}{\dot{\mu}_2 \mu_0}} \dot{R}_{\perp\text{refr}} \cos\theta_2$$

$$(b) \; (1 + \dot{R}_{\parallel}) \cos\theta_1 = \dot{R}_{\parallel\text{refr}} \cos\theta_2 \qquad (7.8)$$

$$\sqrt{\frac{\dot{\epsilon}_1 \epsilon_0}{\dot{\mu}_1 \mu_0}}(1 - \dot{R}_{\parallel}) = \sqrt{\frac{\dot{\epsilon}_2 \epsilon_0}{\dot{\mu}_2 \mu_0}} \dot{R}_{\perp\text{refr}}$$

where $\dot{R}_{\perp}$, $\dot{R}_{\perp\text{refr}}$ and $\dot{R}_{\parallel}$, $\dot{R}_{\parallel\text{refr}}$ are the complex coefficients of reflection and refraction for the two polarizations of the field **E**. Complex Fresnel coefficients of reflection $\dot{R}_{\perp}$ and $\dot{R}_{\parallel}$ can be found from (7.8) using (7.7):

$$\dot{R}_\perp = (\dot{\eta}_\perp \cos\theta_1 - \sqrt{1 - \dot{\nu}^2 \sin^2\theta_1})/(\dot{\eta}_\perp \cos\theta_1 + \sqrt{1 - \dot{\nu}^2 \sin^2\theta_1})$$
(7.9)

$$\dot{R}_\parallel = -(\dot{\eta}_\parallel \cos\theta_1 - \sqrt{1 - \dot{\nu}^2 \sin^2\theta_1})/(\dot{\eta}_\parallel \cos\theta_1 + \sqrt{1 - \dot{\nu}^2 \sin^2\theta_1})$$
(7.10)

where $\dot{\eta}_\perp = \sqrt{\dot{\mu}_2 \dot{\epsilon}_1 / \dot{\mu}_1 \dot{\epsilon}_2}$, and $\dot{\eta}_\parallel = \sqrt{\dot{\mu}_1 \dot{\epsilon}_2 / \dot{\mu}_2 \dot{\epsilon}_1}$.

The Case of Two Layers, of Air and Conductor ($\dot{\mu}_1 = 1$, $\dot{\epsilon}_1 = 1$, $\dot{\epsilon}_2 = \epsilon_2' - j60\lambda\sigma$). It appears frequently in idealized form where the admittance $\sigma \to \infty$, so that $\dot{R}_\perp = \dot{R}_\parallel = -1$. The whole density of the surface electric current $\mathbf{n} \times \mathbf{H} = 2\mathbf{n} \times \mathbf{H}_1$ corresponds then to the double tangential projection of magnetic field $\mathbf{H}_1$ onto interface. The so-called density of "magnetic" current is equal to zero, $\mathbf{n} \times \mathbf{E} = 0$, for the idealized case ($\sigma \to \infty$, $\dot{R}_\perp = \dot{R}_\parallel = -1$), and it is not equal to zero, $\mathbf{n} \times \mathbf{E} \neq 0$, for some real cases. The larger the value of σ, the smaller the depth of penetration δ of the field in conductor (skin depth).

The Case of (N + 1) Layers. Using the ray description (Figure 7.3), Snell's law (7.7), and the well-known equation $e^{\pm j\alpha} = \cos\alpha \pm j\sin\alpha$, we consider the electric and magnetic fields in the nth layer tangential to the interfaces as the superposition of traveling waves or of standing waves in more general form than in [1]:

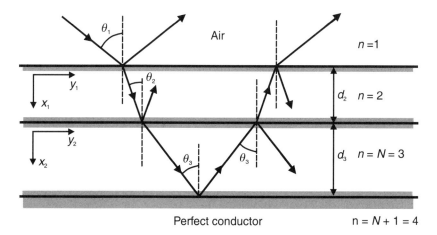

Figure 7.3 Reflection and refraction of waves at interfaces of several infinite dielectric layers (*After:* [1, Figure 8.4]).

$$\left\| \begin{matrix} E_n(x_n) \\ H_n(x_n) \end{matrix} \right\| = \left\| \begin{matrix} A_n e^{-j\alpha_n} + B_n e^{j\alpha_n} \\ Y_n(A_n e^{-j\alpha_n} - B_n e^{j\alpha_n}) \end{matrix} \right\|$$

$$= \left\| \begin{matrix} (A_n + B_n)\cos\alpha_n - (A_n - B_n)j\sin\alpha_n \\ -(A_n + B_n)jY_n\sin\alpha_n + (A_n - B_n)Y_n\cos\alpha_n \end{matrix} \right\| \quad (7.11)$$

where x_n is the coordinate $0 \le x_n \le d_n$ measured from the beginning of the nth layer in the direction normal to the interfaces, A_n and B_n are the amplitudes of the traveling waves in layer n (n = 2, 3, ..., N), $\alpha_n = k_n x_n \cos\theta_n$ is the phase delay, and Y_n is the wave admittance of the layer. For the first n = 0 and the last $n = (N + 1)$ layers the amplitudes of traveling waves are considered only at the interfaces $x_1 = x_{N+1} = 0$, so that $\alpha_1 = \alpha_{N+1} = 0$. The wave admittance is $Y_n = Y_{n\perp} = Y_{n0}\cos\theta_n$ for the E field polarized perpendicularly to the plane of wave incidence and $Y_n = Y_{n\parallel} = Y_{n0}/\cos\theta_n$ for the E field polarized in parallel to the plane of incidence. Here, $Y_{n0} = \sqrt{\epsilon_n\epsilon_0/\mu_n\mu_0}$ is the wave admittance of the layer's medium.

Using (7.11), we introduce the matrices of the field transformation

$$\mathbf{D}(\alpha_n, Y_n) = \left\| \begin{matrix} \cos\alpha_n & -j\sin\alpha_n \\ -jY_n\sin\alpha_n & Y_n\cos\alpha_n \end{matrix} \right\| \quad (7.12)$$

and obtain initial equations describing the field transformation by the nth media layer:

$$\left\| \begin{matrix} E_n(d_n) \\ H_n(d_n) \end{matrix} \right\| = \mathbf{D}(k_n d_n \cos\theta_n, Y_n) \left\| \begin{matrix} A_n + B_n \\ A_n - B_n \end{matrix} \right\|, \quad (7.13)$$

$$\left\| \begin{matrix} E_n(0) \\ H_n(0) \end{matrix} \right\| = \mathbf{D}(0, Y_n) \left\| \begin{matrix} A_n + B_n \\ A_n - B_n \end{matrix} \right\|$$

Using (7.13) and the boundary conditions on the interface $E_n(0) = E_{n-1}(d_{n-1})$, $H_n(0) = H_{n-1}(d_{n-1})$, we obtain the resultant equation of the field transformation by the nth layer

$$\left\| \begin{matrix} E_n(d_n) \\ H_n(d_n) \end{matrix} \right\| = \mathbf{F}(k_n d_n \cos\theta_n, Y_n) \left\| \begin{matrix} E_{n-1}(d_{n-1}) \\ H_{n-1}(d_{n-1}) \end{matrix} \right\| \quad (7.14)$$

where $\mathbf{F}(k_n d_n \cos\theta_n, Y_n) = \mathbf{D}(k_n d_n \cos\theta_n, Y_n)\mathbf{D}^{-1}(0, Y_n)$ or in a more general form

$$\mathbf{F}(\alpha, Y) = \begin{Vmatrix} \cos\alpha & -jY^{-1}\sin\alpha \\ -jY\sin\alpha & \cos\alpha \end{Vmatrix} \tag{7.15}$$

Using (7.14) for the layers $n = N$, $N - 1$, ..., 2 sequentially, we obtain the law of transformation of the tangential components of fields by the set of layers (interfaces)

$$\begin{Vmatrix} E_N(d_N) \\ H_N(d_N) \end{Vmatrix} = \mathbf{F} \begin{Vmatrix} E_1(d_1) \\ H_1(d_1) \end{Vmatrix} = \mathbf{F} \begin{Vmatrix} A_1 + B_1 \\ Y_1(A_1 - B_1) \end{Vmatrix} \tag{7.16}$$

where $\mathbf{F} = \mathbf{F}(k_n d_n \cos\theta_n, Y_n) \cdot \mathbf{F}(k_{n-1}d_{n-1}\cos\theta_{n-1}, Y_{n-1}) \cdot \ldots \cdot \mathbf{F}(k_2 d_2 \cos\theta_2, Y_2)$.

The Case of Layers of "Air—(N – 1) Absorbing Media—Perfect Conductor."
The first layer $n = 1$ is the air, the last layer $n = N + 1$ is a perfect conductor; the interim layers are of dielectric media with absorption, so that the values μ, $\acute{e}$ are complex. All this necessitates an extended interpretation of Snell's law justifying the formal use of complex angles θ to account for the presence of absorption. The layer structure of RAM allows us to augment the domains of model applicability in frequency and angle [1]. On the interface of the Nth layer with the perfect conductor we obtain $E_N(d_N) = E_{N+1}(0) = 0$. Assuming that the coefficient $A_1 = 1$, we find that coefficient B_1 will be equal to the complex coefficient of reflection $\dot{R}$. Then, matrix equation (7.16) of dimension 2×2 takes the form

$$\mathbf{F} = \begin{Vmatrix} F_{11} & F_{12} \\ F_{21} & F_{22} \end{Vmatrix} \begin{Vmatrix} 1 + \dot{R} \\ Y_1(1 - \dot{R}) \end{Vmatrix} = \begin{Vmatrix} 0 \\ H_N(d_N) \end{Vmatrix}$$

Where from $F_{11}(1 + \dot{R}) + F_{12}Y_1(1 - \dot{R}) = 0$, the complex coefficient of reflection will be

$$\dot{R} = -\frac{F_{11} + F_{12}Y_1}{F_{11} - F_{12}Y_1} \tag{7.17}$$

where F_{12} and F_{11} are the elements of matrix $\mathbf{F}$.

The Case of Two Absorbing Layers. The nonabsorbing layer of air ($n = 1$), absorbing layers ($n = 2, 3$), and nonabsorbing layer of perfect conductor ($n = N + 1 = 4$) are considered in this case, and

$$\begin{Vmatrix} F_{11} & F_{12} \\ F_{21} & F_{22} \end{Vmatrix} = \begin{Vmatrix} \cos\alpha_3 & -jY_3^{-1}\sin\alpha_3 \\ -jY_3\sin\alpha_3 & \cos\alpha_3 \end{Vmatrix} \cdot \begin{Vmatrix} \cos\alpha_2 & -jY_2^{-1}\sin\alpha_2 \\ -jY_2\sin\alpha_2 & \cos\alpha_2 \end{Vmatrix}$$

After multiplying the matrices, we obtain $F_{11} = \cos\alpha_2\cos\alpha_3 - Y_2 Y_3^{-1}\sin\alpha_2\sin\alpha_3$ and $F_{12} = -j(Y_2^{-1}\sin\alpha_2\cos\alpha_3 + Y_3^{-1}\cos\alpha_2\sin\alpha_3)$.

Example of Two Thin Absorbing Layers. In this simplest example ($\alpha_2 < 1$, $\alpha_3 < 1$) we have

$$F_{11} \approx 1, \; F_{12}Y_1 \approx -j(Y_1 Y_2^{-1}\alpha_2 + Y_1 Y_3^{-1}\alpha_3),$$
$$\dot{R} \approx -1 + 2j(Y_1 Y_2^{-1}\alpha_2 + Y_1 Y_3^{-1}\alpha_3)$$

If $\alpha_2 = 0$, $\alpha_3 = 0$, the value $|\dot{R}| = 1$. The absorbing layers have negative imaginary parts of the parameters α_2, α_3 leading to a decrease in the modulus of reflection coefficient $|\dot{R}|$.

7.1.4 The Scattered Fields of Huygens Elementary Radiators in Approximation of Physical Optics

Let us return to the fields (7.1) $\mathbf{H}_{tg}(\boldsymbol{\rho})$ and $\mathbf{E}_{tg}(\boldsymbol{\rho}) = \sqrt{\dfrac{\mu_0}{\epsilon_0}}\mathbf{H}_{tg}(\boldsymbol{\rho}) \times \mathbf{R}^0$ of the incident wave on the illuminated part of the target surface. For the cases shown in Figures 7.2 and 7.3 and $\dot{\mu}_1 = \dot{\epsilon}_1 = 1$, we find similarly to (7.5) that

$$\mathbf{n} \times \mathbf{H}_{extr} = \mathbf{n} \times \mathbf{H}_{tg}(\boldsymbol{\rho})[1 - \dot{R}(\boldsymbol{\rho})], \tag{7.18}$$

$$\mathbf{n} \times \mathbf{E}_{extr} = \sqrt{\dfrac{\mu_0}{\epsilon_0}}\mathbf{n} \times [\mathbf{H}_{tg}(\boldsymbol{\rho}) \times \mathbf{R}^0][1 + \dot{R}(\boldsymbol{\rho})]$$

The reflection coefficients $\dot{R}(\boldsymbol{\rho}) = \dot{R}_\perp(\boldsymbol{\rho})$ or $\dot{R}(\boldsymbol{\rho}) = \dot{R}_\parallel(\boldsymbol{\rho})$) depend on wave polarization and the angles of incidence, as was shown in Section 7.1.3.

After applying (7.1), (7.5), (7.6), and (7.18) to the target, being large compared to the wavelength λ, we obtain

$$\mathbf{H}_{rec} \approx jk\int_S \mathbf{B}(\boldsymbol{\rho})\exp(-jkL(\boldsymbol{\rho}))dS, \; \mathbf{E}_{rec} = \sqrt{\dfrac{\mu_0}{\epsilon_0}}\mathbf{H}_{rec} \times \mathbf{r}^0 \tag{7.19}$$

where $\mathbf{B}(\boldsymbol{\rho})$ is the vector-magnitude of radiation of the surface element dS with the radius vector $\boldsymbol{\rho}$

$$\mathbf{B}(\boldsymbol{\rho}) = \frac{1}{4\pi r} \mathbf{J}_c(\mathbf{n}, \mathbf{r}^0) \times \mathbf{r}^0 \tag{7.20}$$

Here, $\mathbf{J}_c(\mathbf{n}, \boldsymbol{\rho})$ is the product of (7.6) for the Huygens elementary radiator and of the multiplier $\exp(jk\boldsymbol{\rho}^{\mathrm{T}}\mathbf{R}^0)$,

$$\mathbf{J}_c(\mathbf{n}, \mathbf{r}^0) = \mathbf{n} \times \mathbf{H}_1 - \sqrt{\frac{\epsilon_0}{\mu_0}} [\mathbf{n} \times \mathbf{E}_1] \times \mathbf{r}^0 \tag{7.21}$$

and $L(\boldsymbol{\rho})$ is the range sum for the element of the surface

$$L(\boldsymbol{\rho}) = r + R + (\mathbf{R}^0 - \mathbf{r}^0)^{\mathrm{T}} \boldsymbol{\rho} \tag{7.22}$$

The first and the second terms of (7.22) describe the range sum to the origin of the target coordinate system. The third term describes the distance from the origin to the corresponding scattering element along the normal to the ellipse of constant range sum [Figure 7.1(b)]. For the monostatic case, $\mathbf{r}^0 = -\mathbf{R}^0$.

Continuing to use approximation of the augmented variant of physical optics, we have

$$\mathbf{H}_1 = \mathbf{H}_0[1 - \dot{R}(\boldsymbol{\rho})], \quad \mathbf{E}_1 = [\mathbf{H}_0 \times \mathbf{R}^0][1 + \dot{R}(\boldsymbol{\rho})]$$

where $\dot{R}(\boldsymbol{\rho})$ is the coefficient of reflection, and $\mathbf{R}^0$ is the unit vector in the direction from transmitting antenna to the target.

7.1.5 The Facet Method of Calculating the Surface Integral and "Cubature" Formulas

The facet method of calculating the surface integral Γ assumes replacing the surface S by the large number plane triangles Δ_i, $i = 1, 2, \ldots, N$ [Figure 7.4(a)], so that

$$\Gamma = \iint_S f(\mathbf{r}) dS \approx \sum_{i=1}^{N} \Gamma_{\Delta i} = \sum_{i=1}^{N} \iint_{\Delta i} f(\mathbf{r}) dS \tag{7.23}$$

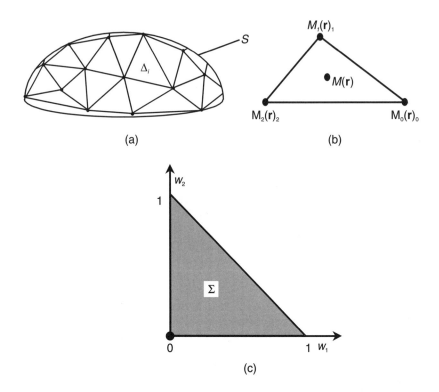

Figure 7.4 The triangles in the facet method of calculating the surface integral: (a) set of triangles, (b) arbitrary triangle, and (c) standard triangle.

Let us consider the integral Γ_Δ over the plane triangle with vertices M_0, M_1, M_2, the radius vectors of which in 3D space are $\mathbf{r}_0$, $\mathbf{r}_1$, $\mathbf{r}_2$ [Figure 7.4(b)]. We describe the position of an arbitrary point M with radius-vector $\mathbf{r}$ by means of its so-called "barycentric" coordinates. The barycentric coordinates are determined as the nonnegative "dot masses" w_0, w_1, w_2 ($w_0 + w_1 + w_2 = 1$) that being disposed at the triangle vertices M_0, M_1, M_2 have the center of mass in the point

$$\mathbf{r} = w_1\mathbf{r}_1 + w_2\mathbf{r}_2 + w_0\mathbf{r}_0 = w_1\boldsymbol{\rho}_1 + w_2\boldsymbol{\rho}_2 + \mathbf{r}_0 \qquad (7.24)$$

Here $\boldsymbol{\rho}_1 = \mathbf{r}_1 - \mathbf{r}_0$, $\boldsymbol{\rho}_2 = \mathbf{r}_2 - \mathbf{r}_0$. The considered integral Γ_Δ, expressed through the barycentric variables w_1, w_2, may be put in the following form

$$\Gamma_\Delta = \int\int_\Sigma f(w_1\boldsymbol{\rho}_1 + w_2\boldsymbol{\rho}_2 + \mathbf{r}_0)\left|\frac{\partial\mathbf{r}}{\partial w_1} \times \frac{\partial\mathbf{r}}{\partial w_2}\right| dw_1 dw_2 \quad (7.25)$$

$$= |\boldsymbol{\rho}_1 \times \boldsymbol{\rho}_2| \int\int_\Sigma \bar{f}[w_1, w_2] dw_1 dw_2$$

where

$$\bar{f}[w_1, w_2] = f(w_1\boldsymbol{\rho}_1 + w_2\boldsymbol{\rho}_2 + \mathbf{r}_0) \quad (7.26)$$

and Σ is the area of integration.

Taking into account that $|\boldsymbol{\rho}_1 \times \boldsymbol{\rho}_2| = 2S_\Delta$ (S_Δ is the area of the triangle Δ), we obtain

$$\Gamma_\Delta = 2S_\Delta \int\int_\Sigma \bar{f}[w_1, w_2] dw_1 dw_2 = 2S_\Delta \int_0^1 dw_1 \int_0^{1-w_1} \bar{f}[w_1, w_2] dw_2$$

$$(7.27)$$

Thus, the calculation of the integrals Γ_Δ over an arbitrary triangle area Δ is reduced to evaluation of the integral over area Σ of a standard triangle [Figure 7.4(c)] on the plane $Ow_1 w_2$ with vertices $(0,0)$, $(0,1)$, $(1,0)$.

For slow varying functions (7.26) and small-sized standard triangle Δ the integral (7.27) can be replaced by the product of the area S_Δ of the triangle and the average value of function $\bar{f}[w_1, w_2]$ at its vertices $(0,0)$, $(0,1)$, (1.0)

$$\Gamma_\Delta = S_\Delta(\bar{f}[1, 0] + \bar{f}[0, 1] + \bar{f}[0, 0])/3 \quad (7.28)$$

which can be regarded as a result of linear approximation of a slowly varying function (7.26).

Unlike the quadrature formulas for calculating the areas in two-dimensional space, the formulas for calculating the volumes in three dimensions are known as "cubature" formulas [9]. Such terminology was preserved in [10] for calculating surface integrals in 3D space. The cubature formulas (7.28) are applicable only for integrand functions varying slowly.

The cubature formulas for oscillating rapidly integrand functions [10, 11] have to provide calculation of integrals

$$\Gamma_\Delta = 2S_\Delta \int\!\!\int_\Sigma e^{jk_0\tilde\Phi[w_1,w_2]}\, \tilde{f}[w_1,\, w_2]\, dw_1 dw_2 \qquad (7.29)$$

where

$$\tilde\Phi[w_1, w_2] = \Phi[w_1(\mathbf{r}_1 - \mathbf{r}_0) + w_2(\mathbf{r}_2 - \mathbf{r}_0) + \mathbf{r}_0]$$

Designating $u = k_0(\tilde\Phi[1,\, 0] - \tilde\Phi[0,\, 0])$, $v = k_0(\tilde\Phi[0,\, 1] - \tilde\Phi[0,\, 0])$, we obtain finally the cubature formulas

$$\Gamma_\Delta \approx \exp(jk_0\tilde\Phi[0,\, 0])\{(\tilde{f}[1,\, 0] - \tilde{f}[0,\, 0])\Gamma_{10} \qquad (7.30)$$
$$+ (\tilde{f}[0,\, 1] - \tilde{f}[0,\, 0])\Gamma_{01} + \tilde{f}[0,\, 0]\Gamma_{00}\}$$

with three elements

$$\Gamma_{lm} = \int\!\!\int_\Sigma e^{j(uw_1 + vw_2)}\, w_1^l w_2^m\, dw_1 dw_2 \ (l;\ m = 1;0,\ 0;1,\ 0;0) \quad (7.31)$$

Introducing the function $\varphi(x) = (\exp(jx) - 1)/jx$, we have

$$\Gamma_{10} = -(\varphi(v) - \varphi(u) - (v - u)\varphi'(u))/(v - u)^2,$$
$$\Gamma_{01} = -(\varphi(u) - \varphi(v) - (u - v)\varphi'(v))/(u - v)^2,$$
$$\Gamma_{00} = (\varphi(u) - \varphi(v))/(j(u - v))$$

7.1.6 Example of RCS Calculation of Targets Uncovered and Covered with RAM for Small Bistatic Angles

To test the cubature formulas, a computer simulation of backscattering was carried out [11]. Three tentative models of a large-sized aircraft with reduced RCS were simulated. All models had wingspan about 50m and fuselage length about 20m. The models had surfaces (1) conducting perfectly and uncovered [Figure 7.5(a)], (2) completely covered by absorber, and (3) partially covered by absorber. The latter case is shown in Figure 7.5(b) where covered areas are marked gray.

Tentative values of the wideband absorber parameters were chosen as follows: thickness—5 cm; relative permittivity and permeability—$\epsilon' = \mu' = 1 - 10i$. These values are typical of the Sommerfeld-type absorber [12].

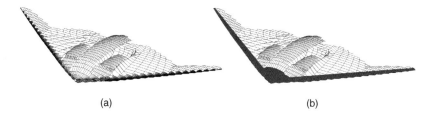

(a)   (b)

Figure 7.5 Aircraft models with reduced RCS and the surface (a) uncovered and (b) partially covered by absorber. The covered areas of surface are marked gray.

Two frequency bands were investigated: 2.25 to 3.75 GHz and 0.15 to 0.25 GHz. A number of triangles amounting to 10^5 provided robustness of measured RCS to the change of the triangle set.

The calculations were carried out using equalities (1.2) for the RCS and (7.19) for the scattered fields. Figure 7.6 shows the calculated average RCS:

- Versus the target course-aspect angle for monostatic radar for zero values of pitch- and roll-aspect angles;
- Versus the radar bistatic angle for zero values of the course-, pitch-, and roll-aspect angles of target illumination.

The RCS of all the models considered was significantly reduced in comparison with usual large-sized aircraft. Especially great reduction of RCS was achieved due to reshaping of target. Covering the surface with RAM, including incomplete coverage, was also effective.

In the frequency band of 2.25 to 3.75 GHz, the averaged monostatic RCS of the uncovered model was reduced up to a 0.1 m^2, but rapidly increased for angles 35° to 40° to about 100 m^2. The latter was because the direction of incidence became normal to the wing edge. In a bistatic radar the RCS increases rapidly at bistatic angles of 70° to 80°.

For the model completely covered with absorber, the maximums of RCS were reduced by two orders of magnitude in comparison with the uncovered model. At some aspect angles the values of RCS were reduced to several hundredths of a square meter.

For the model partially covered with RAM, the maximums of RCS were reduced only by one order of magnitude compared to the uncovered model due to nonuniformity of RAM coating.

In the frequency band of 0.15 to 0.25 GHz the averaged monostatic RCS of the uncovered model was reduced to 1 m^2, but rapidly increased

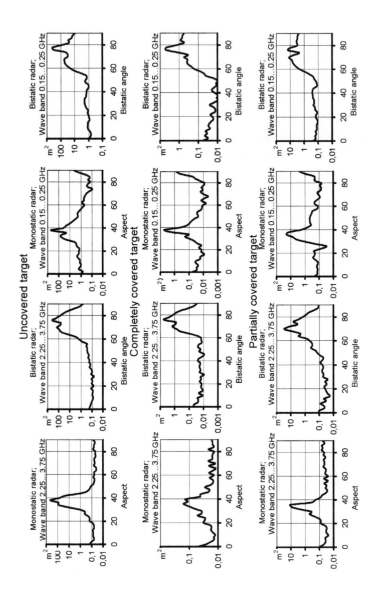

Figure 7.6 Reduced RCS values in two frequency bands calculated for models of large aircraft with uncovered, completely covered, and partially covered surfaces, and given for monostatic radar via target's course-aspect angle and for bistatic radar via bistatic angle for zero course-aspect angle. Zero pitch- and roll-aspect angles were assumed.

to 100 m^2 and even higher for aspect angles 35° to 40°. For the models partially and completely covered with a thick layer of RAM, the RCS maximums were reduced by one to two orders of magnitude compared to the uncovered model.

7.1.7 Evaluation of RCS of Opaque Objects for Bistatic Angles Approaching 180°

Objects with significantly reduced RCS, as well as conducting objects, are usually opaque. An illuminated opaque object with the cross-range dimensions $d \gg \lambda$ forms a shadow region. When illuminating this object by a plane wave of short wavelength, the shadow region becomes a form of a shadow column [6, 12]. The length of the shadow column is about the Fresnel zone of extent d^2/λ. The longitudinal dimensions of an object that is small compared with d^2/λ are assumed to be not very significant for formation of the shadow column. An arbitrary opaque object with such dimensions can be replaced very approximately by the opaque plate that is parallel to the front of incident wave with the surface limited by the column generatrices. In turn, the shadowed part of the opaque plate can be replaced by extraneous electric [$\mathbf{n} \times \mathbf{H}_{extr}$] and "magnetic" [$\mathbf{n} \times \mathbf{H}_{rxtr}$] currents, which together approximately constitute unidirectional Huygens radiators (Section 7.1.4) with a cardioid pattern. Interfering with the incident wave, this radiator forms the shadow column, in which the phases of incident and additional waves near the target must be opposite. Hence, [$\mathbf{R}^0 \times \mathbf{H}_{extr}$] = $-[\mathbf{R}^0 \times \mathbf{H}_0]$ and [$\mathbf{R}^0 \times \mathbf{E}_{extr}$] = $-[\mathbf{R}^0 \times \mathbf{E}_0]$, where $\mathbf{R}^0$, $\mathbf{H}_0$, $\mathbf{E}_0$ were introduced in (7.1). Calculating the fields (7.19) at the point of reception and using (1.2), we can then obtain the corresponding target's RCS

$$\sigma_{tg} = \frac{4\pi A^2}{\lambda^2} F(\beta, \epsilon)$$

Here, A is the area of the shadow as viewed by the transmitter; β and ϵ are the azimuth and elevation angles, measured away from the forward scatter direction; and $F(\beta, \epsilon)$ is a function describing directivity of the forward radiation. The RCS sidelobes can be described by the function $F(\beta, 0) \approx \text{sinc}^2(\pi L \beta/\lambda)$ for a rectangular opaque plate, or by the function $F(\beta, 0) \approx [2J_1(\pi L \beta/\lambda)/(\pi L \beta/\lambda)]^2$ for an ellipsoidal form of a plate, where L is the horizontal dimension of the target across the line of sight. Analogous dependencies exist for the angle coordinate ϵ and the combination of β and ϵ [13].

If $A = 100$ m^2, $\lambda = 0.1$, and $\beta = 0$, the value of $\sigma_{tg} \approx 1.2 \times 10^7$ m^2 for a perfectly plane plate. Hence, a large value of RCS can exist for this approximation also in a very narrow sector of bistatic angles near 180°. In an arbitrary sector of bistatic angles, the RCS diminishes significantly, but it can still be significantly greater than the RCS for small and zero bistatic angles. Design formulas of the type described above must be used carefully, since the sidelobe structure depends on actual distribution of Huygens radiators.

The question arises as to whether it is justifiable to replace an arbitrary convex opaque object by a plate. The answer is positive. It is true that for the illuminated part of the convex object surface, the path length-differences for the illumination and backscattered waves are summed. But for the shadowed part of the convex object surface, such range differences are subtracted. In the whole, (7.19) is applicable both for illuminated and shadowed regions of the object surface provided that $\mathbf{H}_{extr}$ and $\mathbf{E}_{extr}$ were estimated correctly.

7.1.8   Principles of Calculation of RCS for Sharp-Cornered Objects Uncovered and Covered with RAM

We consider the scattering of a plane electromagnetic wave (7.1) by a perfectly conducting object with an absorbing coating of surface discontinuities that can be typical of wings and some other parts covered with RAM. The model of a discontinuity (Figure 7.7) has a form of wedge with an external angle $\pi\gamma$ and a curved edge. The integration surface S is the sum of two surfaces $S = S_0 + S_1$. To evaluate the integral (7.19) over the surface S_1 we can use

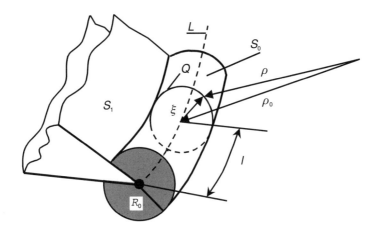

Figure 7.7 The model of a curved wedge coated with RAM.

the facet method (Section 7.1.5). Such evaluation for the surface S_0 is more complicated. But it is simplified by the fact that the radius R_0 of the toroidal absorbing coating is small and the radius R_L of the edge curvature (not shown in Figure 7.7) is large compared to the wavelength λ.

The radius vector $\boldsymbol{\rho}$ of a point on the surface S_0 can be represented as $\boldsymbol{\rho} = \boldsymbol{\rho}_0(l) + \boldsymbol{\xi}(R_0, 0, \varphi)$ (Figure 7.7). Here $\boldsymbol{\rho}_0(l)$ is the radius-vector of a point with position l on the fracture L, and $\boldsymbol{\xi}(R_0, 0, \varphi)$ is the vector of constant length R_0, orthogonal to the curved edge at this point and constituting an angle φ ($0 \leq \varphi \leq \pi\gamma$) with one of the wedge faces, where $\gamma = \gamma(l)$.

According to (7.19) and (7.22) the integral over the surface S_0 can be represented as

$$\mathbf{H}_0 \approx jk \int_L \exp[-jk\, L_L(l)] \int_Q \mathbf{B}(\boldsymbol{\rho}) \exp[-jk\, L_Q(\boldsymbol{\rho})]dqdl \qquad (7.32)$$

$$= jk \int_L \mathbf{M}(l) \exp[-jk\, L_L(l)]dl$$

Here, Q is an arc of the circumference located in the plane orthogonal to L and $dq = R_0 d\varphi$ is the element of its length. In turn, $L_L(l) = (\mathbf{R}^0 - \mathbf{r}^0)^T \boldsymbol{\rho}_0(l)$ and $L_Q(\boldsymbol{\rho}) = -(\mathbf{r}^0)^T \cdot \boldsymbol{\xi}(R_0, \varphi) + \mathrm{const}$ are the range sums depending on l and φ, respectively.

Let us first evaluate the inner integral $\mathbf{M}(l)$ over the arc Q as a function of l. Because of the condition $R_0 < \lambda$, physical optics cannot be used for this purpose. Using the condition $R_L \gg \lambda$, let us consider the asymptotic case $R_L \to \infty$. The problem will be solved for this case as was the cylindrical one by the method of eigenfunctions [14]. Evaluation of the external integral over the arc L under the condition $R_L \gg \lambda$ is facilitated by the presence of a rapidly oscillating multiplier in integrand function and applicability of the method of stationary phase [15].

Solution of the Cylindrical Problem. Let us use the cylindrical coordinate system $OR\varphi z$ matched with the Cartesian one, $Oxyz$, which will be used for description of the exciting plane wave. The systems have the common origin O and axis Oz. The reference line for the angle coordinate φ in the cross section orthogonal to the line L corresponds to the axis Ox. The solution must be found for the air ($n = 1$) and the dielectric ($n = 2$) limited by perfect conductor ($n = 3$) and air ($n = 1$). The solution can be found as a superposition of the waves in the waveguides ($n = 1, 2$) of electric E

(transverse magnetic TM) and magnetic H (transverse electric TE) types of various modes coupled together by the boundary conditions and character of exciting plane wave. Solutions for the axial components of field have the form

$$
\left\| \begin{matrix} E_z^{(n)}(R,\, \varphi,\, z) \\ H_z^{(n)}(R,\, \varphi,\, z) \end{matrix} \right\| = \tag{7.33}
$$

$$
e^{-j\alpha z} \sum_{m=0}^{\infty} \left\| \begin{matrix} [A_m^{(n)} J_{\nu(m)}(\kappa^{(n)}R) + B_m^{(n)} H_{\nu(m)}^{(1)}(\kappa^{(n)}R)]\, \sin[\nu(m)\varphi] \\ [C_m^{(n)} J_{\nu(m)}(\kappa^{(n)}R) + D_m^{(n)} H_{\nu(m)}^{(1)}(\kappa^{(n)}R)]\, \cos[\nu(m)\varphi] \end{matrix} \right\|
$$

Other components can be obtained, as in waveguide theory, from Maxwell equations; for instance,

$$
E_R^{(n)} = -\frac{j\alpha}{[\kappa^{(n)}]^2}\frac{\partial E_z^{(n)}}{\partial R} + \frac{j\omega\mu}{[\kappa^{(n)}]^2 R}\frac{\partial H_z^{(n)}}{\partial\varphi}, \tag{7.34}
$$

$$
E_\varphi^{(n)} = -\frac{j\omega\mu}{[\kappa^{(n)}]^2}\frac{\partial H_z^{(n)}}{\partial R} + \frac{j\alpha}{[\kappa^{(n)}]^2 R}\frac{\partial E_z^{(n)}}{\partial\varphi}
$$

Let us clarify the meaning of this solution and the designations used. Solution of the cylindrical problem for dielectric layers $n = 1,\ 2$ can be considered as a development of the more simple solution (7.11) of the plane problem for such layers $n = 1, 2, \ldots, N$. In addition to solutions (7.11), the solutions (7.33) ($n = 1, 2$) are the superposition of particular solutions of Maxwell's equations (eigenfunctions) in the wave form, but with some peculiarities. They are built as a superposition of infinite number of nonuniform cylindrical waves of the type $E(R,\ \varphi)e^{j(\omega t \pm \alpha z)}$. They are standing waves along the coordinate φ since the components E_z and E_R tangential to the perfectly conducting wedge must vanish at its surface together with the partial derivative $\dfrac{\partial H_z^{(1)}}{\partial\varphi}$ [see (7.34)]. Therefore, $\sin[\nu(m)\varphi] = 0$ for $\varphi = 0$ and $\varphi = \pi\gamma$, so that $\nu(m)\pi\gamma = m\pi$, and

$$
\nu(m) = m/\gamma,\ m = 0, 1, 2, \ldots \tag{7.35}
$$

The radial distribution of the field must be described by the superposition of standing and traveling nonuniform cylindrical waves corresponding to various azimuth distributions $\nu(m)$. The standing wave is described by the Bessel functions $J_{\nu(m)}(\kappa^{(n)}R)$, and the traveling wave is described by the Hankel functions $H_{\nu(m)}^{(1)}(\kappa^{(n)}R)$ and $H_{\nu(m)}^{(2)}(\kappa^{(n)}R)$. For the assumed time dependence $e^{j\omega t}$, the function $H_{\nu(m)}^{(1)}(\kappa^{(n)}R)$ describes a traveling wave of convergent (approaching) type $e^{j(\omega t + \kappa R)}$, and the function $H_{\nu(m)}^{(2)}(\kappa^{(n)}R)$ describes a traveling wave of divergent (receding) type $e^{j(\omega t - \kappa R)}$. The cylindrical standing wave is a superposition of opposite traveling waves

$$J_{\nu(m)}(\kappa^{(n)}R)e^{j\omega t} = \frac{1}{2}[H_{\nu(m)}^{(1)}(\kappa^{(n)}R) + H_{\nu(m)}^{(2)}(\kappa^{(n)}R)]e^{j\omega t} \quad (7.36)$$

analogous to the plane standing wave $\cos kx \, e^{j\omega t} = \frac{1}{2}[e^{j(\omega t + kx)} + e^{j(\omega t - kx)}]$. Asymptotic approximations $z = \kappa^{(n)}R \to \infty$ of these functions are [14]

$$H_{\nu}^{(1)}(z) \approx \sqrt{\frac{2}{\pi z}} \exp\left[j\left(z - \nu\frac{\pi}{2} - \frac{\pi}{4}\right)\right] \quad (7.37)$$

$$H_{\nu}^{(2)}(z) \approx \sqrt{\frac{2}{\pi z}} \exp\left[-j\left(z - \nu\frac{\pi}{2} - \frac{\pi}{4}\right)\right]$$

Using (7.33), (7.36), and (7.37), we can approximate the component E_z in the far-field zone as a superposition of traveling waves convergent and divergent in radial direction:

$$E_z^{(1)}(R, \varphi, z)e^{j\omega t} \approx \sqrt{\frac{2}{\pi \kappa^{(1)}R}} e^{-j\frac{\pi}{4}} \{f_1(\varphi)\exp[j(\omega t + \kappa^{(1)}R + \alpha z)]$$

$$+ f_2(\varphi)\exp[j(\omega t - \kappa^{(1)}R + \alpha z)]\} \quad (7.38)$$

For $\alpha = 0$, both waves are cylindrical ones; for $\alpha \neq 0$, they are conical ones. Functions $f_1(\varphi)$ and $f_2(\varphi)$ describe angular distributions of convergent and divergent waves, so

$$f_1(\varphi) = \sum_{m=0}^{\infty}(A_m^{(1)}/2 + B_m^{(1)})e^{-j[\nu(m)\pi/2]}\sin[\nu(m)\varphi] = \int_0^{\gamma\pi} f_1(\psi)\delta(\psi - \varphi)d\psi$$

$$(7.39)$$

Let us compare the equations obtained with the equation of a plane wave propagating from a remote source disposed in direction φ_0, θ_0 from the object considered:

$$E_z(t, x, y, z) = A \exp[j(\omega t + k_1 x + k_2 y + k_3 z)] = E_z(t, R, \varphi, \theta) \quad (7.40)$$

$$= E_{z0} \exp[j(\omega t + \kappa R + k_3 z)]$$

where $k_1 = \dfrac{\omega}{c} \cos\varphi_0 \sin\theta_0$, $\quad k_2 = \dfrac{\omega}{c} \sin\varphi_0 \sin\theta_0$, $\quad k_3 = \dfrac{\omega}{c} \cos\theta_0$,

$R = \sqrt{x^2 + y^2}$ and $\kappa R = \sqrt{(k_1 x)^2 + (k_2 y)^2}$, so that $\kappa = \dfrac{\omega}{c} \sin\theta_0$. For the plane wave polarized in the incidence plane passing through the axis z with intensity of electric field E, the value $E_{z0} = E \cos\theta$. It is equal to zero in the case of polarization normal to this plane.

Equations (7.38) and (7.40) can be brought into accord only if $\kappa^{(1)} = \kappa$ and $\alpha = k_3$. Assuming this, we conclude that the general solution (7.39) corresponds also to the superposition of the plane waves converging to the object from various directions φ. Our single plane wave corresponds to the function $f_1(\varphi) = E_{z0} \delta(\varphi - \varphi_0)$. This delta-function can be considered as a limit for $R \to \infty$ of the function

$$d(\varphi, R, a) = R/a \text{ if } |\varphi - \varphi_0| < a/2R, \text{ and}$$

$$d(\varphi, R, a) = 0 \text{ if } |\varphi - \varphi_0| > a/2R$$

provided that additional suppositions are made of $R_0 \ll a \ll R$, so that the product of the maximum value of the function R/a and the length of interval a/R of its nonzero values is equal to unity. The additional supposition of $R_0 \ll a \ll R$ ensures the plane wave-front in the neighborhood of the object's discontinuity (Figure 7.8).

The condition of agreement of (7.38), (7.39), and (7.40) for $R \to \infty$ is

$$\sqrt{\frac{2}{\pi\kappa^{(1)}R}}\, e^{-j\frac{\pi}{4}} \sum_{m=0}^{\infty} (A_m^{(1)}/2 + B_m^{(1)}) e^{-j[\nu(m)\pi/2]} \sin[\nu(m)\varphi] \approx E_{z0}\, d(\varphi, R, a)$$

$$(7.41)$$

It allows us to express the linear combination $A_m^{(1)}/2 + B_m^{(1)}$ of the parameters of the cylindrical wave in air ($n = 1$) with a given intensity E_{z0} as the coefficients of Fourier transformation (7.41) for the angle interval

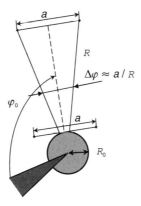

Figure 7.8 Clarification of asymptotic $(R \to \infty)$ decomposition of the cylindrical (conical) nonuniform waves on the plane uniform waves.

$\gamma\pi$. Similarly, the linear combination $C_m^{(1)}/2 + D_m^{(1)}$ can be expressed through the given component H_{z0}.

Using all these coefficients, let us return to (7.33) including the Bessel and Hankel functions connecting the fields in air ($n = 1$) and dielectric ($n = 2$) media.

It is convenient to describe the fields in dielectric media ($n = 2$) only by Bessel functions since for $R \to \infty$ each of the Hankel functions $H_{\nu(m)}^{(1)}(\kappa^{(n)}R) \to \infty$ and $H_{\nu(m)}^{(2)}(\kappa^{(n)}R) \to \infty$. Therefore, $B^{(2)} = D^{(2)} = 0$. Parameters $\kappa^{(n)}$, identical for the Bessel and Hankel functions, can be found according to Helmholz equation

$$(\Delta + \omega^2 \mu^{(n)} \epsilon^{(n)})E_z = 0$$

Applying this equation to each of the partial solutions of (7.33), we have

$$-(\kappa^{(n)})^2 - \alpha^2 + \omega^2 \mu^{(n)} \epsilon^{(n)} = 0 \text{ and } \kappa^{(n)} = \sqrt{\omega^2 \mu^{(n)} \epsilon^{(n)} - \alpha^2}$$

$$(7.42)$$

The value of $\kappa^{(1)}$ obtained from (7.42) for air corresponds to the value obtained above.

The coefficients $A_m^{(2)}$ and $C_m^{(2)}$ can be found from the boundary conditions $E_z^{(2)} = E_z^{(1)}, E_\varphi^{(2)} = E_\varphi^{(1)}, H_z^{(2)} = H_z^{(1)}, H_\varphi^{(2)} = H_\varphi^{(1)}$. Due to orthogonality of the basis functions of the Fourier transform, the boundary conditions will be simplified. For example, the first of these conditions takes the form

$$A_m^{(2)} J_{\nu(m)}(\kappa^{(2)} R_0) = A_m^{(1)} J_{\nu(m)}(\kappa^{(1)} R_0) + B_m^{(1)} H_{\nu(m)}^{(1)}(\kappa^{(1)} R_0)$$

The boundary condition for E_φ can be obtained using equality (7.34). Analogous equations can be written for the components H_z and H_φ. Together with the known values of linear combinations $A_m^{(1)}/2 + B_m^{(1)}$ and $C_m^{(1)}/2 + D_m^{(1)}$ for each value m we have six linear equations to obtain six unknown parameters: $A_m^{(1)}$, $B_m^{(1)}$, $C_m^{(1)}$, $D_m^{(1)}$, $A_m^{(2)}$, $C_m^{(2)}$. This solves the problems of the field evaluation. Then, using the methods of numerical integration, we can calculate the inner integral $\mathbf{M}(l)$ of equality (7.32).

Application of the Method of Stationary Phase. The integral (7.32) as a whole

$$\mathbf{H}_0 = jk \int_L \mathbf{M}(l) \exp[-jk L_L(l)] dl \qquad (7.43)$$

has as an integrand function the product of the rapidly oscillating function

$$\exp[-jk L_L(\boldsymbol{\rho}_0)] = \cos[-jk L_L(\boldsymbol{\rho}_0)] + j \sin[-jk L_L(\boldsymbol{\rho}_0)]$$

and the internal integral $\mathbf{M}(l)$ of (7.32), as a function of l, that varies slowly. The areas of positive and negative half-waves of (7.43) compensate each other except in the neighborhood of the points of "stationary phase" (specular bright points) l_i, $i = 1, 2, \ldots, N$, where the derivative of the range sum $L_L(l)$ is equal to zero

$$\partial L_L(l_i)/\partial l = 0, \quad L_L(l) = (\mathbf{R}^0 - \mathbf{r}^0)^{\mathrm{T}} \boldsymbol{\rho}_0(l)$$

Let us approximate the radius-vectors $\boldsymbol{\rho}_0(l)$ of the points of a small part of the curve neighboring the stationary phase point l_i by three first terms of the Taylor series

$$\boldsymbol{\rho}_0(l) \approx \boldsymbol{\rho}_0(l_i) + \frac{\partial \boldsymbol{\rho}_0(l_i)}{\partial l}(l - l_i) + \frac{1}{2} \frac{\partial^2 \boldsymbol{\rho}_0(l_i)}{\partial l^2}(l - l_i)^2$$

The same part of the curve L can be also approximated by the arc of circumference. The first derivative $\partial \boldsymbol{\rho}_0(l_i)/\partial l$ is defined, then, as the limit of the ratio of the vectors' difference $\Delta \boldsymbol{\rho}_0 = \boldsymbol{\rho}_0(l_i + \Delta l/2) - \boldsymbol{\rho}_0(l_i - \Delta l/2)$, corresponding to the arc subtense to the length of the arc Δl for $\Delta l \to 0$ [Figure 7.9(a)]. The value of this derivative $\partial \boldsymbol{\rho}_0(l_i)/\partial l$ is equal to the unit vector $\boldsymbol{\tau}_i = \boldsymbol{\tau}(l_i)$ of the tangent to the curve L at the point [Figure7.9(a)].

The second derivative $\partial^2 \boldsymbol{\rho}_0(l_i)/\partial l^2 = \partial \boldsymbol{\tau}(l_i)/\partial l$ can be defined as the limit of the ratio of the vectors' difference $\Delta \boldsymbol{\tau} = \boldsymbol{\tau}(l_i + \Delta l/2) - \boldsymbol{\tau}(l_i - \Delta l/2)$ corresponding to some subtense of the arc of unit radius [Figure 7.9(b)] to the length of other arc $\Delta l = R_{Li}\Delta\beta$ of the radius R_{Li} for $\Delta l \to 0$. The difference $\Delta \boldsymbol{\tau}$ can be evaluated as $\Delta \boldsymbol{\tau} = \boldsymbol{\nu}_i \Delta\beta$, where $\boldsymbol{\nu}_i$ is the unit vector

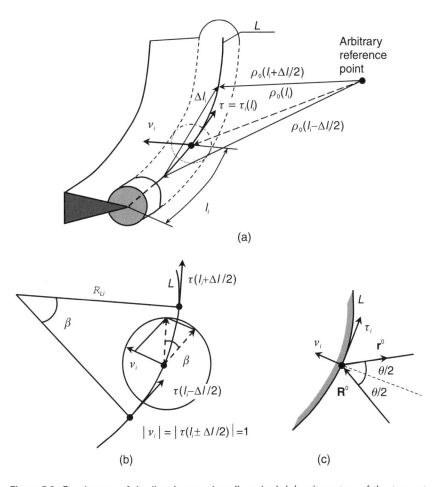

(a)

(b)   (c)

Figure 7.9 For the use of the "stationary phase" method: (a) unit vectors of the tangent and normal to the convex curve L (Figure 7.8) and clarification of the value $\partial \boldsymbol{\rho}_0(l_i)/\partial l$; (b) clarification of the value $\partial^2 \boldsymbol{\rho}_0(l_i)/\partial l^2$; and (c) unit vectors of wave propagation at the "stationary phase" point for bistatic radar.

of the principal normal to the convex curve L at the point l_i. Therefore, $\partial^2 \boldsymbol{\rho}_0(l_i)/\partial l^2 \approx \boldsymbol{\nu}_i/R_{Li}$, and finally

$$\boldsymbol{\rho}_0(l) \approx \boldsymbol{\rho}_0(l_i) + (l - l_i)\boldsymbol{\tau}_i + \frac{\boldsymbol{\nu}_i}{2R_{Li}}(l - l_i)^2 \qquad (7.44)$$

The condition of the phase stationarity at the point $l = l_i$ is

$$\frac{\partial L_L(l)}{\partial l} = (\mathbf{R}^0 - \mathbf{r}^0)^{\mathrm{T}}\frac{\partial}{\partial l}\boldsymbol{\rho}_0(l) = (\mathbf{R}^0 - \mathbf{r}^0)^{\mathrm{T}}\boldsymbol{\tau}(l) = 0 \text{ for } l = l_i$$

$$(7.45)$$

which corresponds to the bisector of bistatic angle orthogonal to the tangent to the curve L at this point [Figure 7.9(c)].

In the neighborhood of the point $l = l_i$, the value of the range sum is equal to

$$L_L(l) = (\mathbf{R}^0 - \mathbf{r}^0)^{\mathrm{T}}\boldsymbol{\rho}_0(l) \approx L_L(l_i) + \zeta_i(l - l_i)^2/2R_{Li} \qquad (7.46)$$

Here $L_L(l_i) = (\mathbf{R}^0 - \mathbf{r}^0)^{\mathrm{T}}\boldsymbol{\rho}_0(l_i)$ and $\zeta_i = (\mathbf{R}^0 - \mathbf{r}^0)^{\mathrm{T}}\boldsymbol{\nu}_i = |\zeta_i|\operatorname{sgn}\zeta_i$, where $\operatorname{sgn}\zeta_i = 1$ if $\zeta_i > 0$ (the convex surfaces and curves), and $\operatorname{sgn}\zeta_i = -1$ if $\zeta_i < 0$ (the concave ones). Using the tabulated integral

$$\int_{-\infty}^{\infty} e^{j\frac{\pi}{2}u^2}\, du = \sqrt{2}e^{j\frac{\pi}{4}}$$

approximation (7.46), and designating $\dfrac{k|\zeta_i|(l - l_i)^2}{2R_{Li}} = \dfrac{\pi}{2}u^2$, so that $dl = \sqrt{\dfrac{\pi R_{Li}}{k|\zeta_i|}}\, du$, we obtain the contribution of a "visible" stationary phase (specular) point into integral (7.43)

$$\mathbf{H}_{0i} \approx \sqrt{\frac{2\pi R_{Li}}{k|\zeta_i|}}\,\mathbf{M}(l_i)\exp\left[j\left(-kT_L(l_i) + \frac{\pi}{4}\operatorname{sgn}\zeta_i\right)\right] \qquad (7.47)$$

Adding together the contributions of all "visible" specular points into integral (7.43), we obtain the resultant field $\mathbf{H}_0$. The sum of $\mathbf{H}_0$ and $\mathbf{H}_1$, where $\mathbf{H}_1$ is the integral through the surface S_1 computed by the use of the facet method (Section 7.1.5), gives the requested field $\mathbf{H}_{rec}$.

Let us note that we need not integrate only through the surfaces S_0 and S_1 shown in Figure 7.7. For the convenience of calculations, one may integrate through the surface S_0' embracing S_0 and a residuary part of the surface S_1 where the current distribution is more uniform. Examples of calculation and comparison of its results with experimental data were presented in [14, 15].

7.2 Some Calculating Methods for Nonstationary Illumination of Targets

We consider below the concept of the high frequency impulse and unit step responses of targets (Section 7.2.1), and the method (Section 7.2.2) and examples (Sections 7.2.3 and 7.2.4) of calculating such responses and transient response of a wideband signal. Let us limit the discussion here only to targets with perfectly conducting surfaces [16–18].

Before developing the calculating methods, let us consider some physical features of scattered signals in bistatic radar systems for extended targets. Considering monostatic radar in previous chapters, we operated with the RPs of targets. The bistatic radar directly measures not the range of a target or its elements but the corresponding range sums. The RPs of monostatic radar can be replaced, therefore, by the range sum profiles (RSP) of bistatic radar. The first element of the RP corresponds to the target element illuminated first. The first element of the RSP corresponds to the target element with minimum range sum, which may not be illuminated first if it is nearer to the receiver.

7.2.1 Concept of High Frequency Responses of Targets

The theory of linear systems operates with the impulse response (IR) and unit step response (USR). This approach can be applied for calculating both the IRs and USRs, but some notes are necessary.

The impulse response is one of the most important characteristics of linear systems, especially those with constant parameters. The IR is defined as the output for $t \geq 0$ of the linear system exposed to the action of the Dirac delta-function applied at its input at the moment $t = 0$. The convolution of any given signal with the IR gives the system response to this signal. In experimental and theoretical investigations, the delta-pulse can be replaced by a short pulse having the same amplitude-phase spectrum in a definite frequency region. The convolution rule will be correct then only for the signals in this frequency region.

Unit step response is a reaction of a linear system to the unit step, being the integral of the Dirac delta-function. One can calculate the response to a given signal by convolution of its time derivative with USR of the system.

High-frequency impulse responses (HFIRs) and high-frequency unit step responses (HFUSRs) are obtained by the formal using of the delta-functions and unit step functions in the high-frequency domain. The necessity of introducing the separate responses for this domain arises from the absence of general calculating methods suitable for simulating the scattering for the high and not very high frequencies. The physical optics approximation of HFIR was used, for instance, in the work [16] by the name of impulse response.

7.2.2 Calculating Bistatic Responses of Targets with Perfectly Conducting Surfaces Using the Physical Optics Approach

Using (7.22), let us introduce first the path-delay sum for various elements of the target surface described by the vector ρ in the target coordinate system

$$T(\rho) = L(\rho)/c = [(\mathbf{R}^0 - \mathbf{r}^0)^T \rho + r + R]/c \qquad (7.48)$$

The common IR of a linear system is the Fourier transform of its amplitude-phase response. Using (7.19) as the amplitude-phase response of a target to a sinusoid of arbitrary frequency f and replacing the k by its value $\dfrac{2\pi f}{c}$, we obtain the HFIR in the form

$$H(t) \approx \int\limits_{-\infty}^{\infty} j\frac{2\pi f}{c} \int\limits_{S_1} \mathbf{B}(\rho) \exp[j2\pi f[t - T(\rho)]]dSdf \qquad (7.49)$$

$$= \int\limits_{S_1} \mathbf{B}(\rho) \int\limits_{-\infty}^{\infty} \frac{\partial}{\partial t} \exp[j2\pi f[t - T(\rho)]]dfdS$$

The relations for HFIR $H(t)$ and HFUSR $\mathbf{H}(t)$ can be presented in the form

$$H(t) \approx \frac{\partial}{\partial t} \int_{S_1} \delta[t - T(\boldsymbol{\rho})]\mathbf{B}(\boldsymbol{\rho})dS = \frac{\partial}{\partial t}\mathbf{H}(t), \qquad (7.50)$$

where $\mathbf{H}(t) \approx \int_{S_1} \delta[t - T(\boldsymbol{\rho})]\mathbf{B}(\boldsymbol{\rho})dS$

since the delta-function and unit step function are defined as

$$\delta(\tau) = \int_{-\infty}^{\infty} \exp(j2\pi f\tau)df \quad \text{and} \quad \chi(t) = \int_{0}^{t} \delta(s)ds \qquad (7.51)$$

The second of expressions (7.50) can be rewritten in two equivalent forms

$$H(t) \approx \frac{\partial}{\partial t} \int_{S_1} \chi[t - T(\boldsymbol{\rho})]\mathbf{B}(\boldsymbol{\rho})dS = \frac{\partial}{\partial t} \int_{S_t} \mathbf{B}(\boldsymbol{\rho})dS \qquad (7.52)$$

where S_t is the part of the surface S_1 that had been illuminated to the moment $t = T(\boldsymbol{\rho})$.

Introducing the auxiliary vector-function of time

$$\mathbf{N} = \mathbf{N}(t) = \int_{S_t} \mathbf{n}dS \qquad (7.53)$$

where $S_t = S_1(t)$, we can obtain its first derivative

$$\frac{\partial \mathbf{N}}{\partial t} = \lim_{\Delta t \to 0} \frac{1}{\Delta t}\left[\int_{S_{t+\Delta t}} \mathbf{n}dS - \int_{S_t} \mathbf{n}dS \right] = \lim_{\Delta t \to 0} \frac{1}{\Delta t} \int_{S_{t+\Delta t} \setminus S_t} \mathbf{n}dS \qquad (7.54)$$

Designation $S_{t+\Delta t} \setminus S_t$ in (7.54) corresponds to integration through the part of the surface $S_{t+\Delta t}$, which does not belong to the surface S_t. Assuming that the electrical field tangential to the surface is equal to zero, and using (7.19) through (7.21) and (7.53) and (7.54), we obtain the HFUSR and HFIR in the form

$$\mathbf{H}(t) \approx \left[\frac{\partial \mathbf{N}}{\partial t} \times \tilde{\mathbf{H}}_0 \right] \times \mathbf{r}^0, \ H(t) \approx \left[\frac{\partial^2 \mathbf{N}}{\partial t^2} \times \tilde{\mathbf{H}}_0 \right] \times \mathbf{r}^0 \qquad (7.55)$$

where $\tilde{\mathbf{H}}_0 = \dfrac{1}{4\pi r} \cdot 2\mathbf{H}_{tg}(0)$. As it follows from (7.52), the time boundaries of nonzero values of HFIR exist:

$$t_{min} = \min_{\rho \in S_1} T(\boldsymbol{\rho}), \ t_{max} = \max_{\rho \in S_1} T(\boldsymbol{\rho})$$

The Responses of Targets to Arbitrary High Frequency Signals. The HFIR and HFUSR can be used to obtain the responses of targets on the signal $U_f(t)$ with derivative $U_f'(t)$:

$$\mathbf{Z}(t) = \int_{-\infty}^{\infty} U_f(s) H(t - s) ds, \ \ \mathbf{Z}(t) = \int_{-\infty}^{\infty} U_f'(s) H(t - s) ds \qquad (7.56)$$

The second variant of computations (7.56) is preferable since it requires only the evaluation of the first derivative $\partial \mathbf{N}/\partial t$ in (7.55). Substituting the first of expressions (7.55) in (7.56), we obtain

$$\mathbf{Z}(t) = [\mathbf{Z}_a(t) \times \tilde{\mathbf{H}}_0] \times \mathbf{r}^0 \qquad (7.57)$$

where $\mathbf{Z}_a(t)$ is an auxiliary vector-function

$$\mathbf{Z}_a(t) = \int_{-\infty}^{\infty} U_f'(s) \frac{\partial \mathbf{N}(t - s)}{\partial t} ds \qquad (7.58)$$

For smooth targets and signals with the time-stationary polarization, the time dependence (7.58) completely defines the time dependence (7.57).

Peculiarities of Calculating the Derivative of Vector-Function $\mathbf{N}(t)$. Let us introduce:

- Gradient $\mathbf{grad}\, T(\boldsymbol{\rho}) = \nabla T(\boldsymbol{\rho}) = (\mathbf{R}^0 - \mathbf{r}^0)/c$ of path-delay function (7.48) and its unit vector $\mathbf{m}^0 = (\mathbf{R}^0 - \mathbf{r}^0)/|\mathbf{R}^0 - \mathbf{r}^0|$ oriented along the bisector of the bistatic angle;

- Contour $\Gamma(t)$, being the intersection of the plane $cT(\boldsymbol{\rho}) = ct$ with the surface S_1;
- Differential dl of this contour's length.

Returning to expression (7.53), we obtain

$$dS = dl \cdot c\Delta t/\sin(\mathbf{m}^0, \mathbf{n}) = c \cdot dl \cdot \Delta t/\sqrt{1 - \cos^2(\mathbf{m}^0, \mathbf{n})} \quad (7.59)$$

$$= c \cdot dl \cdot \Delta t/\sqrt{1 - [(\mathbf{m}^0)^{\mathrm{T}}\mathbf{n}]^2}$$

so that, according to (7.54),

$$\frac{\partial \mathbf{N}}{\partial t} = c \int\limits_{\Gamma(t)} \frac{\mathbf{n}}{\sqrt{1 - [(\mathbf{m}^0)^{\mathrm{T}}\mathbf{n}]^2}} dl \quad (7.60)$$

7.2.3 Example of Calculating the HFUSR of Ellipsoids with Perfectly Conducting Surfaces

The surface S of an ellipsoid satisfies the equation

$$\left|A^{-1}\mathbf{x}\right|^2 = \mathbf{x}^{\mathrm{T}}A^{-2}\mathbf{x} = 1, \quad A = A^{\mathrm{T}} = \mathrm{diag}(a_1, a_2, a_3) \quad (7.61)$$

The vector of normal $\mathbf{n}$ to this surface is collinear to the gradient $\nabla(\mathbf{x}^{\mathrm{T}}A^{-2}\mathbf{x}) = 2A^{-2}\mathbf{x}$.

In our succeeding consideration, the concept of the terminator introduced for monochromatic illumination will be useful because it determines the surface S_1. The terminator can be found from the conditions that (1) it belongs to the ellipsoid surface and (2) it consists of the points in which the unit vector $\mathbf{R}^0$ becomes a tangent to the surface orthogonal to its normal $\mathbf{n}$. The terminator can be represented, therefore, by the system of equations

$$\left|A^{-1}\mathbf{x}\right|^2 = 1, \quad \mathbf{n}^{\mathrm{T}}\mathbf{R}^0 = \mathbf{x}^{\mathrm{T}}A^{-2}\mathbf{R}^0 = 0 \quad (7.62)$$

The problem of scattering of an ellipsoid can be reduced to that of a sphere using the coordinate transformations $A^{-1}\mathbf{x} = \boldsymbol{\xi}$ and $A^{-1}\mathbf{R}^0 = \mathbf{Q}$, so that

$$\left|\boldsymbol{\xi}\right|^2 = 1, \quad \mathbf{Q}^{\mathrm{T}}\boldsymbol{\xi} = 0 \qquad (7.63)$$

Figure 7.10 explains the method of computation of sphere scattering in bistatic radar. After the coordinate transformation the accessory vector $\mathbf{Q}$ replaces the unit vector $\mathbf{R}^0$ in the direction of incident wave propagation, the vector

$$\mathbf{L}^0 = \mathbf{Am}^0 / \left|\mathbf{Am}^0\right| \qquad (7.64)$$

replaces the unit vector $\mathbf{m}^0 = (\mathbf{R}^0 - \mathbf{r}^0)/\left|\mathbf{R}^0 - \mathbf{r}^0\right|$ in the direction of propagation of a plane of constant range sum. The integration in (7.60) will be carried out over the moving contour $\Gamma_1(t')$ instead of the moving contour $\Gamma(t)$. Here, t' is the relative time measured with respect to the moment of the contour $\Gamma(t)$ passing through the center of ellipsoid. It is normalized in relation to the time $\left|\mathbf{A}(\mathbf{R}^0 - \mathbf{r}^0)\right|/c$ of propagating the contour $\Gamma(t)$ from the point of tangency D to the center of ellipsoid

$$t' = \frac{t - (r + R)/c}{\left|\mathbf{A}(\mathbf{R}^0 - \mathbf{r}^0)\right|/c} = \frac{ct - (r + R)}{\left|\mathbf{A}(\mathbf{R}^0 - \mathbf{r}^0)\right|}$$

We will denote the values of $t'_{\min}$ and $t'_{\max}$ as corresponding to the values of $t_{\min}$ and $t_{\max}$ introduced above. The motion of the contour $\Gamma_1(t')$

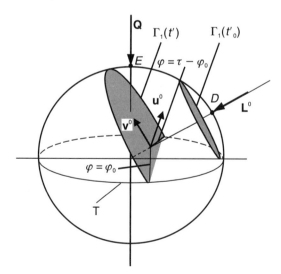

Figure 7.10 Geometry of the nonstationary problem of sphere scattering in bistatic radar.

in the direction $\mathbf{L}^0$ along the sphere is shown in Figure 7.10. It begins from the point D, where the plane of constant range sum is a tangent to the sphere. The echo signal corresponding to the point D is received earlier than that corresponding to the point E, despite the fact that the point E was illuminated earlier. The latter is due to the smaller distance from the point D to receiver. The contour $\Gamma_1(t')$ is a full circumference until it is truncated by the terminator plane $(t' < t'_0)$, otherwise $\Gamma_1(t')$ is a truncated circumference $(t'_0 < t' < t'_{max})$.

To calculate the contour integral, we introduce as auxiliary variables the angular coordinate φ and two vectors, $\mathbf{u}^0 = -\mathbf{L}^0 \times \mathbf{Q}/|\mathbf{L}^0 \times \mathbf{Q}|$ that are parallel to the terminator plane and $\mathbf{v}^0 = \mathbf{u}^0 \times \mathbf{L}^0$. Both vectors are disposed in the plane of contour $\Gamma_1(t')$ (Figure 7.10). Then,

$$\boldsymbol{\xi} = t'\mathbf{L}^0 + \sqrt{1 - (t')^2}\,\mathbf{F}(\varphi), \quad \mathbf{F}(\varphi) = \mathbf{u}^0\cos\varphi + \mathbf{v}^0\sin\varphi, \quad (7.65)$$

$$\varphi_0(t') \le \varphi \le \pi - \varphi_0(t')$$

Here, the $t'\mathbf{L}^0$ is the vector normal to the plane of contour $\Gamma_1(t')$. Its length $|t'|$ is equal to the distance from this plane to the origin of coordinate system. The length $\sqrt{1 - (t')^2}$ is equal to the radius of the circumference including the contour $\Gamma_1(t')$. These results correspond to a sphere of unit radius and zero time reference connected with the contour plane passing through the center of the sphere.

The value $\varphi_0(t')$ entered in (7.65) can be defined from the condition of intersection (Figure 7.10) of the terminator plane $\mathbf{Q}^T\boldsymbol{\xi} = 0$ and the contour $\Gamma_1(t')$ defined by (7.62), so that

$$\mathbf{Q}^T\boldsymbol{\xi} = \mathbf{Q}^T[t'\mathbf{L}^0 + \sqrt{1 - (t')^2}(\mathbf{u}^0\cos\varphi_0 + \mathbf{v}^0\sin\varphi_0)] = 0$$

Since $\mathbf{Q}^T\mathbf{u}^0 = 0$, we have $\varphi_0(t') = \mathrm{asin}(t'/\alpha\sqrt{1 - (t')^2})$, where $\alpha = \mathbf{Q}^T\mathbf{v}^0/\mathbf{Q}^T\mathbf{L}^0$, if $t'_0 \le t' \le t'_{max}$ and $\varphi_0(t') = -\pi/2$ if $-1 \le t' < t'_0$. The values $t'_{min} = -1$, $t'_{max} = \alpha/\sqrt{1 + \alpha^2}$ $(\alpha \ge 0)$, and $t'_0 = -t'_{max}$.

Returning to coordinates $\mathbf{x} = \mathbf{A}\boldsymbol{\xi}$, we can obtain parametric equation of the contour $\Gamma(t)$ on the ellipsoid

$$\mathbf{x} = \mathbf{A}\boldsymbol{\xi} = t'\mathbf{A}\mathbf{L}^0 + \sqrt{1 - (t')^2}\,\mathbf{A}\mathbf{F}(\varphi), \quad \varphi_0(t') \le \varphi \le \pi - \varphi_0(t')$$

$$(7.66)$$

The contour (7.66) is the contour of integration to be used in (7.54). Let us return to (7.54) and introduce (1) the relation for unit vector normal to the ellipsoid at points of contour $\Gamma(t)$:

$$\mathbf{n} = \mathbf{A}^{-1}(t'\mathbf{L}^0 + \sqrt{1 - (t')^2}\,\mathbf{F}(\varphi)) / \left| \mathbf{A}^{-1}(t'\mathbf{L}^0 + \sqrt{1 - (t')^2}\,\mathbf{F}(\varphi)) \right| \tag{7.67}$$

and (2) the relation between dl and $d\varphi$:

$$dl = \sqrt{1 - (t')^2} \left| \mathbf{A}\mathbf{F}'(\varphi) \right| d\varphi \tag{7.68}$$

We can then rewrite expression (7.54) in the form

$$\frac{\partial \mathbf{N}}{\partial t} = \int_{\varphi_0(t')}^{\pi - \varphi_0(t')} \mathbf{f}(\varphi, t') d\varphi \tag{7.69}$$

where

$$\mathbf{f}(\varphi, t') = G(\varphi)\mathbf{A}^{-1}(t'\mathbf{L}^0 + \sqrt{1 - (t')^2}\,\mathbf{F}(\varphi)) \tag{7.70}$$

$$G(\varphi) = \left| \mathbf{A}\mathbf{F}'(\varphi) \right| / (\left| \mathbf{A}\mathbf{F}'(\varphi) \right|^2 - ((\mathbf{m}^0)^{\mathrm{T}}[\mathbf{A}^{-1}\mathbf{F}])^2)^{1/2} \tag{7.71}$$

The relation (7.69) can be represented in the form

$$\frac{\partial \mathbf{N}}{\partial t} = \tag{7.72}$$

$$\left[\chi(t' + 1) \int_0^{2\pi} \mathbf{f}(\varphi, t') d\varphi - \chi(t' - t'_0) \int_{\pi - \varphi_0(t')}^{2\pi + \varphi_0(t')} \mathbf{f}(\varphi, t') d\varphi \right] \chi(t'_{\max} - t')$$

where $\chi(t')$ is the unit step function. This result can be used to obtain HFUSR of ellipsoid from the second of these relations (7.50). The HFIR of ellipsoid can be obtained from the first of these relations (7.50) [17, 18].

7.2.4 Example of Calculating the Transient Response of an Aircraft Model with Conducting Surface for a Wideband Signal

Let us consider an aircraft model (Figure 7.11) containing several triaxial ellipsoids I_1, I_2, ..., I_m ($m = 5$). Calculation of the value $\dfrac{\partial \mathbf{N}}{\partial t}$ is reduced to the following steps:

- Estimation of values $t_{\min l}$ and $t_{\max l}$ for each ellipsoid ($l = 1, 2, \ldots, m$);
- Calculation of radius-vectors $\mathbf{x}_l(t)$ of the points corresponding to the interval of polar angles $\varphi_0(t'_l) < \varphi_l < \pi - \varphi_0(t'_l)$ for the contours $\Gamma_l(t)$ belonging to each lth ellipsoid;
- Test of each point $\mathbf{x}_l(t)$ for visibility (or shadowing) [19];
- Calculation of contour integral $\Gamma(t)$ for the whole aircraft model as a function of time t performing integration over separate contours $\Gamma_l(t)$ by use of expression (7.72);
- Calculation of convolution (7.58) as a function of t using Filon's type formula [20].

Transient responses of a simplified aircraft model (Figure 7.11) were calculated for the following sizes of ellipsoids' I_l half-axes:

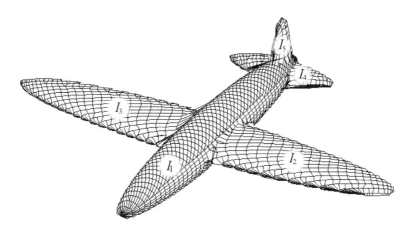

Figure 7.11 The simplified model of an aircraft used in calculation of transient responses without separating the bright elements.

I_1: $a_1 = 1.25$m, $a_2 = 1.25$m, $a_3 = 9$m;

I_2: $a_1 = 0.5$m, $a_2 = 11$m, $a_3 = 2$m;

I_3: $a_1 = 0.5$m, $a_2 = 11$m, $a_3 = 2$m;

I_4: $a_1 = 0.3$m, $a_2 = 3$m, $a_3 = 1$m;

I_5: $a_1 = 3$m, $a_2 = 0.3$m, $a_3 = 1$m

The centers of ellipsoids I_1, I_2, I_3 were placed at the origin of the target coordinate system, and the centers of ellipsoids I_4 and I_5 were at the distance 7.6m from the origin along the longitudinal axis.

The calculated responses (RSPs) are shown in Figure 7.12 for a radio pulse of duration $\tau_p = 3$ ns at carrier frequency 10 GHz. The illumination directions were supposed lying in the wing plane. Responses of Figure 7.12(a) and (b) correspond to the nose-on illumination. Responses of Figure 7.12(c) and (d) correspond to illumination at 30° from the nose. Responses of Figure 7.12(a) and (c) were obtained for bistatic angle of 20°. Responses of Figure 7.4(b) and (d) were obtained for bistatic angle of 40°. One can see that local responses from different parts of the model aircraft are well separated.

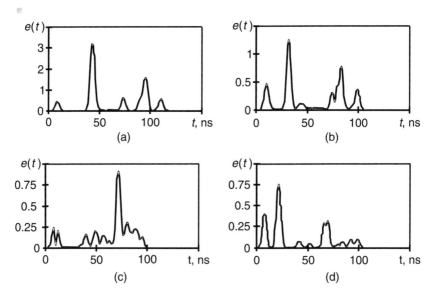

Figure 7.12 The RSPs of aircraft model (Figure 7.11) calculated without separation of bright elements for the signal bandwidth of 300 MHz: (a) course-aspect angle 0° and bistatic angle 20°; (b) course-aspect angle 0° and bistatic angle 40°; (c) course-aspect angle 30° and bistatic angle 20°; and (d) course-aspect angle 30° and bistatic angle 40°.

Details of the images (Figure 7.12) are represented more poorly than those of the images obtained by using the simplest component method (Chapters 1–6) for the same bandwidth (about 300 MHz). This can be explained by the oversimplification of the aircraft model connected with its demonstrative objective and the limited computer capability.

References

[1]   Knott, E. F., J. F. Shaeffer, and M. T. Tuley, *Radar Cross Section*, Second Edition, Norwood, MA: Artech House, 1993.

[2]   Kerr, D. E. (ed.), *Propagation of Short Radio Waves*, MIT Radiation Laboratory Series, No. 13, New York: McGraw-Hill, 1951.

[3]   Silver, S., *Microwave Antenna Theory and Design*, MIT Radiation Laboratory Series, No. 12, New York: McGraw-Hill, 1949.

[4]   Vainshtein, L. A., *Electromagnetic Waves*, Moscow: Sovetskoe Radio Publishing House, 1988 (in Russian).

[5]   Ufimtsev, P. Y., *Method of Edge Waves in the Physical Theory of Diffraction*, Moscow: Sovetskoe Radio Publishing House, 1962 (in Russian). Translated by U.S. Air Force, Foreign Technol. Div. Wright-Patterson AFB, OH, 1971: Tech. Rep. AD N 733203 DTIC. Cameron Station, Alexandria, VA 22304-6145.

[6]   Ufimtsev, P. Y., "Comments on Diffraction Principles and Limitations of RCS Reduction Techniques," *Proc. IEEE*, Vol. 84, December 1996, pp. 1828–1851.

[7]   Fock, V. A, *Problems of Diffraction and Propagation of Electromagnetic Waves*, Moscow: Sovetskoe Radio Publishing House, 1970 (in Russian).

[8]   Sukharevsky, O. I., "Electrodynamic Calculation of the Model of Two-Reflector Antenna with Strict Accounting for the Interaction between the Reflectors," *Radiotekhnika*, Vol. 60, 1982, Kharkov.

[9]   Sobolev, S.L., *Introduction into the Theory of Cubature Formulas*, Moscow: Nauka Publishing House, 1974 (in Russian).

[10]   Zamyatin, V. I., B. N. Bahvalov, and O. I. Sukharevsky, "Computation of Patterns of the Bent Radiating Surfaces," *Radiotekhnika i Electronika*, Vol. 23, June 1978 (in Russian).

[11]   Sukharevsky, O. I. et al., "Calculation of Electromagnetic Wave Scattering on Perfectly Conducting Object Partly Coated by Radar Absorbing Material with the Use of Triangulation Cubature Formula," *Radiophyzika and Radioastronomiya*, Vol. 5, No. 1, 2000 (in Russian).

[12]   Zahariev, L. N., and A. A Lemansky, *The Scattering of Waves by "Black" Bodies*, Moscow, Sovetskoe Radio Publishing House, 1972 (in Russian).

[13]   Barton, D. K., *Modern Radar System Analysis*, Norwood, MA: Artech House, 1988.

[14]   Sukharevsky, O. I., and A. F. Dobrodnyak, "A Three-Dimensional Diffraction Problem on the Perfectly Conducting Wedge with Radio-Absorbing Cylinder on the Edge,"

Izvestiya Vysshih Uchebnyh Zavedeniy USSR, Radiofizika, Vol. 31, September 1988, pp. 1074–1081 (in Russian).

[15] Sukharevsky, O. I., and A. F. Dobrodnyak, "The Scattering by Finite Perfectly Conducting Cylinder with Absorbing Coated Edges in Bistatic Case," *Izvestiya Vysshih Uchebnyh Zavedeniy USSR, Radiofizika*, Vol. 32, December 1989, pp. 1518–1524 (in Russian).

[16] Kennaugh, E. M., and D. L Moffatt, "Transient and Impulse Response Approximations," *Proc. IEEE*, Vol. 53, August 1965, pp. 893–901.

[17] Sukharevsky, O. I., and V. A. Vasilets, "Impulse Characteristics of Smooth Objects in Bistatic Case," *Journal of Electromagnetic Waves and Applications*, Vol. 10, December 1996, pp. 1613–1622.

[18] Sukharevsky, O. I. et al., "Pulse Signal Scattering from Perfectly Conducting Complex Object Located near Uniform Half-Space," *Progress in Electromagnetic Research*, Vol. 29, 2000, pp. 169–185.

[19] Rodgers, D. F., *Procedural Elements for Computer Graphics*, New York: McGraw-Hill, 1985.

[20] Tranter, C. J., *Integral Transforms in Mathematical Physics*, Moscow: Foreign Literature Publishing House, 1956 (in Russian).

List of Acronyms

1D	one-dimensional
2D	two-dimensional
3D	three-dimensional
ADC	analog-to-digital converter
AGC	automatic gain control
ALCM	air launch cruise missile
AN	artificial neuron
ANN	artificial neural network
ANN M	artificial neural network, modularized
ANN NM	artificial neural network, nonmodularized
CFAR	constant false alarm rate
cpdf	conditional probability density function
CW	continuous wave
DFT	discrete Fourier transform
DTM	digital terrain map
ET	evolutionary training
FANN	feedforward artificial neural network
FANN NM	feedforward artificial neural network, nonmodularized
FFT	fast Fourier transform
GLCM	ground launch cruise missile
H	horizontal
HFIR	high frequency impulse response
HFUSR	high frequency unit step response
HRR	high range resolution

IMRQ	information measure of recognition quality
IR	impulse response
ISAR	inverse synthetic aperture radar
JEM	jet-engine modulation
lcpdf	logarithmic conditional probability density function
LFM	linear frequency modulation
MTI	moving target indicator
MTD	moving target detector
NB	narrowband
pdf	probability density function
PGA	pair gradient algorithm
PPI	plan-position indicator
PRF	pulse repetition frequency
PRM	propeller modulation
PSM	polarization scattering matrix
PW	precursory weighting
R&D	research and development
RAM	radar absorbing material
RCS	radar cross-section
RFP	range frequency profile
RMS	rotational modulation signature
RP	range profile
RPP	range polarization profile
RSP	range sum profile
SF	stepped frequency
SGA	simple gradient algorithm
SNR	signal-to-noise ratio
SS	signature set
TFDS	time-frequency distribution series
USR	unit step response
V	vertical
V-H	velocity-altitude
WB	wideband
WE	without engine
WF	wavefront
WRP	wavelet range profile
WV	Wigner-Ville

About the Authors

Yakov D. Shirman (1919) is a professor at the Kharkov Military University, Ukraine. He holds a Ph.D. in radio communications (1948) from Leningrad Airforce Academy; a D.Sc. in radar (1960) from the Institute of Radio Engineering at the Academy of Sciences of USSR, Moscow; a Diploma of Honorable Worker of Science and Technology of Ukraine (1967); and two State Prize Diplomas (1979, 1988). He is the author of many books on electronic systems, their statistical theory, and electrodynamics. His inventions laid the foundation of the work in the areas of pulse compression and adaptive antennas in the former USSR.

Sergey A. Gorshkov (1961) is a professor at the Military Academy, Belarus, Minsk. He holds a Ph.D. in radar (1990) from the Radio Engineering Academy of Air Defense, Kharkov. He is the author of more than 70 scientific works and inventions in the field of target backscattering, radar recognition, and wave propagation.

Sergey P. Leshchenko (1959) is a section head of the Research Center at the Kharkov Military University, Ukraine. He holds a Ph.D. in radar (1992) from the Radio Engineering Academy of Air Defense, Kharkov. He is the author of approximately 50 scientific works in the field of radar recognition.

Valeriy M. Orlenko (1970) is a teacher at the Kharkov Military University, Ukraine. He holds a Ph.D. in radar (1998) from the Kharkov Military University. He is the author of more than 15 works in the field of radar recognition and neural networks.

Sergey Yu. Sedyshev (1961) is a professor at the Military Academy, Belarus, Minsk. He holds a Ph.D. in radar (1991) from the Minsk Antiaircraft Missile Military High School of Air Defense. He is the author of more than 40 scientific works in the field of radar and computer modeling.

Oleg I. Sukharevskiy (1950) is a professor at the Kharkov Military University, Kharkov, Ukraine. He holds a Ph.D. in radar (1983) from the USSR Scientific and Research Institute of Radio Engineering, Moscow, and a D.Sc. in radar (1993) from the Radio Engineering Academy of Air Defense, Kharkov. He is the author of more than 100 scientific works in the field of diffraction theory and other branches of radiophysics.

Index

Multitarget-Multisensor Tracking: Applications and Advances Volume III, Yaakov Bar-Shalom and William Dale Blair, editors

Principles of High-Resolution Radar, August W. Rihaczek

Radar Cross Section, Second Edition, Eugene F. Knott, et al.

Radar Evaluation Handbook, David K. Barton, et al.

Radar Meteorology, Henri Sauvageot

Radar Reflectivity of Land and Sea, Third Edition, Maurice W. Long

Radar Resolution and Complex-Image Analysis, August W. Rihaczek and Stephen J. Hershkowitz

Radar Signal Processing and Adaptive Systems, Ramon Nitzberg

Radar System Performance Modeling, G. Richard Curry

Radar Technology Encyclopedia, David K. Barton and Sergey A. Leonov, editors

Range-Doppler Radar Imaging and Motion Compensation, Jae Sok Son, et al.

Theory and Practice of Radar Target Identification, August W. Rihaczek and Stephen J. Hershkowitz

For further information on these and other Artech House titles, including previously considered out-of-print books now available through our In-Print-Forever® (IPF®) program, contact:

Artech House	Artech House
685 Canton Street	46 Gillingham Street
Norwood, MA 02062	London SW1V 1AH UK
Phone: 781-769-9750	Phone: +44 (0)20 7596-8750
Fax: 781-769-6334	Fax: +44 (0)20 7630-0166
e-mail: artech@artechhouse.com	e-mail: artech-uk@artechhouse.com

Find us on the World Wide Web at:
www.artechhouse.com